The Hidden Process

By

Ian Beardsley

Copyright © 2015 by Ian Beardsley

ISBN: 978-1-329-11197-4

We seem to be part of some sort of a process, just as Robert Dean suggested, but whether it is due to extraterrestrials, or rather some unascertainable force, we cannot rule out that it is due to a natural evolution with nothing behind it.

What we may be unraveling here is the history of the evolution of written languages, whether they be writing, mathematics, physics, or computer science. We may be discovering those first primal sounds that our ancient ancestors first uttered to represent the various aspects of reality.

Here is how it works:

Vea means see in spanish, but sounds the same as bella in spanish means beautiful, and sounds like bea in beautiful, and vela means candle in spanish, but sounds just like bela in italian meaning beautiful. So we write:

Que vea la bea ut por la luz de la vela (That he see the beautiful ut by the light of the candle). The only sound not connected to anything is ut, and we say ut was the first word that was uttered by an prehistoric ancestor that meant woman. This is not necessarily true, but it is an example of the kind of process we use to put together the story of the evolution of spoken languages.

Another example:

The Maidel is Maid for Madam. "Maidel" is Yiddish for "Girl", a Maid cleans the house, is a woman historically, and Madam is a play on these words. Maid, sounds like made, and we write:

woman = made (Adam made Eve from his rib.)

We can write: Man made maidel and called her ut the maid.

Maid = made (That which is made).

In the Gypsy language, Roma, candela means fire, sounds like candle, and in spanish vela means candle sounds just like bela (Italian for beautiful) and we write:

Vela la candela (Both candle the fire and beautiful the fire).

Mira is Spanish for look, mirror means reflect, and we write "Mira in the mirror".

Finally we write: The Roma Rome The Land of Rome (The Gypsies Rome the land of Rome). Keep in mind Gypsies are wanderers.

Let's apply this to computer languages and mathematics. We will see the evolution of our systems of units (those magnitudes we invented to measure things), are connected to Nature even though we did not intend to make them that way. This is what is meant by we are part of a process.

Ian Beardsley
May 03 2015

An Interesting view of numbers and their connection to Nature comes to us from the Gypsy Caves of Southern Spain and results in an interpretation of computer science that not only might connect it to extraterrestrials, but that suggests it might be something other than what we think it is.

Ian Beardsley
May 02 2015

AE-35

I wrote a short story last night, called Gypsy Shamanism and the Universe about the AE-35 unit, which is the unit in the movie and book 2001: A Space Odyssey that HAL reports will fail and discontinue communication to Earth. I decided to read the passage dealing with the event in 2001 and HAL, the ship computer, reports it will fail in within 72 hours. Strange, because Venus is the source of 7.2 in my Neptune equation and represents failure, where Mars represents success.

Ian Beardsley
August 5, 2012

Chapter One

It must have been 1989 or 1990 when I took a leave of absence from The University Of Oregon, studying Spanish, Physics, and working at the state observatory in Oregon -- Pine Mountain Observatory—to pursue flamenco in Spain.

The Moors, who carved caves into the hills for residence when they were building the Alhambra Castle on the hill facing them, abandoned them before the Gypsies, or Roma, had arrived there in Granada Spain. The Gypsies were resourceful enough to stucco and tile the abandoned caves, and take them up for homes.

Living in one such cave owned by a gypsy shaman, was really not a down and out situation, as these homes had plumbing and gas cooking units that ran off bottles of propane. It was really comparable to living in a Native American adobe home in New Mexico.

Of course living in such a place came with responsibilities, and that included watering its gardens. The Shaman told me: "Water the flowers, and, when you are done, roll up the hose and put it in the cave, or it will get stolen". I had studied Castilian Spanish in college and as such a hose is "una manguera", but the Shaman called it "una goma" and goma translates as rubber. Roll up the hose and put it away when you are done with it: good advice!

So, I water the flowers, rollup the hose and put it away. The Shaman comes to the cave the next day and tells me I didn't roll up the hose and put it away, so it got stolen, and that I had to buy him a new one.

He comes by the cave a few days later, wakes me up asks me to accompany him out of The Sacromonte, to some place between there and the old Arabic city, Albaicin, to buy him a new hose.

It wasn't a far walk at all, the equivalent of a few city blocks from the caves. We get to the store, which was a counter facing the street, not one that you could enter. He says to the man behind the counter, give me 5 meters of hose. The man behind the counter pulled off five meters of hose from the spindle, and cut the hose to that length. He stated a value in pesetas, maybe 800, or so, (about eight dollars at the time) and the Shaman told me to give that amount to the man behind the counter, who was Spanish. I paid the man, and we left.

I carried the hose, and the Shaman walked along side me until we arrived at his cave where I was staying. We entered the cave stopped at the walk way between living room and kitchen, and he said: "follow me". We went through a tunnel that had about three chambers in the cave, and entered one on our right as we were heading in, and we stopped and before me was a collection of what I estimated to be fifteen rubber hoses sitting on ground. The Shaman told me to set the one I had just bought him on the floor with the others. I did, and we left the chamber, and he left the cave, and I retreated to a couch in the cave living room.

Chapter Two

Gypsies have a way of knowing things about a person, whether or not one discloses it to them in words, and The Shaman was aware that I not only worked in Astronomy, but that my work in astronomy involved knowing and doing electronics.

So, maybe a week or two after I had bought him a hose, he came to his cave where I was staying, and asked me if I would be able to install an antenna for television at an apartment where his nephew lived.

So this time I was not carrying a hose through The Sacromonte, but an antenna.

There were several of us on the patio, on a hill adjacent to the apartment of The Shaman's Nephew, installing an antenna for television reception.

Chapter Three

I am now in Southern California, at the house of my mother, it is late at night, she is a asleep, and I am about 24 years old and I decide to look out the window, east, across The Atlantic, to Spain. Immediately I see the Shaman, in his living room, where I had eaten a bowl of the Gypsy soup called Puchero, and I hear the word Antenna. I now realize when I installed the antenna, I had become one, and was receiving messages from the Shaman.

The Shaman's Children were flamenco guitarists, and I learned from them, to play the guitar. I am now playing flamenco, with instructions from the shaman to put the gypsy space program into my music. I realize I am not just any antenna, but the AE35 that malfunctioned aboard The Discovery just before it arrived at the planet Jupiter in Arthur C. Clarke's and Stanley Kubrick's "2001: A Space Odyssey". The Shaman tells me, telepathically, that this time the mission won't fail.

Chapter Four

I am watching Star Wars and see a spaceship, which is two oblong capsules flying connected in tandem. The Gypsy Shaman says to me telepathically: "Dios es una idea: son dos". I understand that to mean "God is an idea: there are two elements". So I go through life basing my life on the number two.

Chapter Five

Once one has tasted Spain, that person longs to return. I land in Madrid, Northern Spain, The Capitol. The Spaniards know my destination is Granada, Southern Spain, The Gypsy Neighborhood called The Sacromonte, the caves, and immediately recognize I am under the spell of a Gypsy Shaman, and what is more that I am The AE35 Antenna for The Gypsy Space Program. Flamenco being flamenco, the Spaniards do not undo the spell, but reprogram the instructions for me, the AE35 Antenna, so that when I arrive back in the United States, my flamenco will now state their idea of a space program. It was of course, flamenco being flamenco, an attempt to out-do the Gypsy space program.

Chapter Six

I am back in the United States and I am at the house of my mother, it is night time again, she is asleep, and I look out the window east, across the Atlantic, to Spain, and this time I do not see the living room of the gypsy shaman, but the streets of Madrid at night, and all the people, and the word Jupiter comes to mind and I am about to say of course, Jupiter, and The Spanish interrupt and say "Yes, you are right it is the largest planet in the solar system, you are right to consider it, all else will flow from it."

I know ratios, in mathematics are the most interesting subject, like pi, the ratio of the circumference of a circle to its diameter, and the golden ratio, so I consider the ratio of the orbit of Saturn (the second largest planet in the solar system) to the orbit of Jupiter at their closest approaches to The Sun, and find it is nine-fifths (nine compared to five) which divided out is one point eight (1.8).

I then proceed to the next logical step: not ratios, but proportions. A ratio is this compared to that, but a proportion is this is to that as this is to that. So the question is: Saturn is to Jupiter as what is to what? Of course the answer is as Gold is to Silver. Gold is divine; silver is next down on the list. Of course one does not compare a dozen oranges to a half dozen apples, but a dozen of one to a dozen of the other, if one wants to extract any kind of meaning. But atoms of gold and silver are not measured in dozens, but in moles. So I compared a mole of gold to a mole of silver, and I said no way, it is nine-fifths, and Saturn is indeed to Jupiter as Gold is to Silver.

I said to myself: How far does this go? The Shaman's son once told me he was in love with the moon. So I compared the radius of the sun, the distance from its center to its surface to the lunar orbital radius, the distance from the center of the earth to the center of the moon. It was Nine compared to Five again!

Chapter Seven

I had found 9/5 was at the crux of the Universe, but for every yin there had to be a yang. Nine fifths was one and eight-tenths of the way around a circle. The one took you back to the beginning which left you with 8 tenths. Now go to eight tenths in the other direction, it is 72 degrees of the 360 degrees in a circle. That is the separation between petals on a five-petaled flower, a most popular arrangement. Indeed life is known to have five-fold symmetry, the physical, like snowflakes, six-fold. Do the algorithm of five-fold symmetry in reverse for six-fold symmetry, and you get the yang to the yin of nine-fifths is five-thirds.

Nine-fifths was in the elements gold to silver, Saturn to Jupiter, Sun to moon. Where was five-thirds? Salt of course. "The Salt Of The Earth" is that which is good, just read Shakespeare's "King Lear". Sodium is the metal component to table salt, Potassium is, aside from being an important fertilizer, the substitute for Sodium, as a metal component to make salt substitute. The molar mass of potassium to sodium is five to three, the yang to the yin of nine-fifths, which is gold to silver. But multiply yin with yang, that is nine-fifths with five-thirds, and you get 3, and the earth is the third planet from the sun.

I thought the crux of the universe must be the difference between nine-fifths and five-thirds. I subtracted the two and got two-fifteenths! Two compared to fifteen! I had bought the Shaman his fifteenth rubber hose, and after he made me into the AE35 Antenna one of his first transmissions to me was: "God Is An Idea: There Are Two Elements".

It is so obvious, the most abundant gas in the Earth Atmosphere is Nitrogen, chemical group 15 and the Earth rotates through 15 degrees in one hour.

I have said, since my theory suggest extraterrestrials gave us our units of measurement, that extraterrestrials might have given us our variables used in physics and math, like the unit vectors (i, j, k). I have already found a pattern and posted it. However, I was doing my CS50x computer science homework and trying to write a program for Caesar's Cipher. I wrote a small program and decided to test it. If you write a program and test it, standard input is "hello". I put in hello and to test, ran the program for rotating characters by 1, and 2, and 3, as they are the first integers and the easiest with which to test your program. The result was the "h" on "hello", came out to be (i, j, k). In other words you get that (i, j, k) is a hello from aliens in accordance with my earlier theories. If this is not real contact with extraterrestrials, it is great content for a Sci-Fi movie about contact with extraterrestrials. Here is the program I wrote, and the result of running it:

As you can see I am making some kind of a cipher, but not Caesar's Cipher

```
#include <stdio.h>
#include <cs50.h>
#include <string.h>
int main(int argc, string argv[1])
{
int i=0;
int k = atoi(argv[1]);
if (argc>2 || argc<2)
printf ("Give me a single string: ");
else
printf("Give me a phrase: ");
string s = GetString();
for (int i =0, n=strlen(s); i<n; i++);
printf("%c", s[i]+k);
printf("\n");
}
```

Running Julius 01

```
jharvard@appliance (~): cd Dropbox/pset2
jharvard@appliance (~/Dropbox/pset2): make julius
clang -ggdb3 -O0 -std=c99 -Wall -Werror    julius.c  -lcs50 -lm -o julius
jharvard@appliance (~/Dropbox/pset2): ./julius 3
Give me a phrase: hello
k
jharvard@appliance (~/Dropbox/pset2): ./julius 4
Give me a phrase: hello
l
jharvard@appliance (~/Dropbox/pset2): ./julius 2
Give me a phrase: hello
j
jharvard@appliance (~/Dropbox/pset2): ./julius 1
Give me a phrase: hello
i
jharvard@appliance (~/Dropbox/pset2):
```

I posted to my blog http://cosasbiendichas.blogspot.com/

Sunday, January 26, 2014

A Pattern Emerges

(a, b, c) in ASCII computer code is (97, 98, 99) the first three numbers before a hundred and 100 is totality (100%).

(i, j, k) in numeric are is (9, 10, 11) the first three numbers before twelve and 12 is totality in the sense that 12 is the most abundant number for its size
(divisible by 1,2, 3, 4, 6 = 16) is larger than 12).

(x, y, z) in ASCII computer code is (120, 121, 122) the first three numbers before 123 and 123 is the number with the digits 1, 2, 3 which are the numeric numbers for the
(a, b, c) that we started with.

Thursday, January 23, 2014

We Look Further Into Human Definitions That Seem Arbitrary

Just as we found our units of measurement, what they evolved into being and how we defined them, are centered around the triad of 9/5, 5/3, and 15, we might ask are our common usage of variables connected to Nature and the Universe as well. In pursuing such a question we look at:

$(x, y, z,)$ as they represent the three axis is rectangular coordinates. We look at (i, j, k) as as they are the representations for the unit vectors, and they correspond respectively to (x, y, z). We also look at (n) as it often means "number" and we look at $(p$ and $q)$ as they range from 0 to 1, in probability problems. We might first look at their binary and hexadecimal equivalents to get a start, if not their decimal equivalents. (i) is also often "integer" and (a, b, c) are the coefficients of a quadratic and are the corners of a triangle. We might add that (s) is length, as in physics $dW=F\,ds$. (a, b, c) have the same kind of correspondence with (x, y, z) as (i, j, k). All three sets, then, line up with one another and are at the basis of math and physics.

To learn of my evidence in support of the idea extraterrestrials left their thumbprint in our physics and that they embedded a message in our physics that seems to come from the same region in space as the SETI Wow! Signal, Sagittarius, read my book All That Can Be Said.

rotate all characters in string

```
#include <stdio.h>
#include <cs50.h>
#include <string.h>
int main(int argc, string argv[1])
{

int k = atoi(argv[1]);
if (argc>2 || argc<2)
printf ("Give me a single string: ");
else
printf("Give me a phrase: ");
string s = GetString();
for (int i =0, n=strlen(s); i<n; i++)
{
printf("%c", s[i]+k);
}
printf("\n");
}
```

Both hola and hello work. Hola is Spanish for Hello

khoor
jharvard@appliance (~/Dropbox/psets/pset2): ./caesar 1
Give me a phrase: hola
ipmb
jharvard@appliance (~/Dropbox/psets/pset2): ./caesar 2
Give me a phrase: hola
jqnc
jharvard@appliance (~/Dropbox/psets/pset2): ./caesar 3
Give me a phrase: hola
krod
jharvard@appliance (~/Dropbox/psets/pset2):

Let's take out the index i in the array and put in 0 to look at the first letter in the string

```c
#include <stdio.h>
#include <cs50.h>
#include <string.h>
int main(int argc, string argv[1])
{
int k = atoi(argv[1]);
if (argc>2 || argc<2)
printf ("Give me a single string: ");
else
printf("Give me a phrase: ");
string s = GetString();
for (int i =0, n=strlen(s); i<n; i++)
printf("%c", s[0]+k);
printf("\n");
}
```

```
jharvard@appliance (~): cd Dropbox/psets/pset2
jharvard@appliance (~/Dropbox/psets/pset2): make julius
clang -ggdb3 -O0 -std=c99 -Wall -Werror    julius.c  -lcs50 -lm -o julius
jharvard@appliance (~/Dropbox/psets/pset2): ./julius 1
Give me a phrase: hello
iiiii
jharvard@appliance (~/Dropbox/psets/pset2): ./julius 2
Give me a phrase: hello
jjjjj
jharvard@appliance (~/Dropbox/psets/pset2): ./julius 3
Give me a phrase: hello
kkkkk
```

the program enquiry.c

```c
#include <stdio.h>
int main(void)
{
int i=3;
int n[i];
for (int i=0; i<3;i++)
{
printf("%d ET-X, acknowledge enquiry: ", i);
scanf("%d", &n[i]);
}
if (n[0]==9 && n[1]==10 && n[2]==11)
{
printf("hello\n");
}
else
{
printf("Not the right response.");
}
}
```

running enquiry.c

```
jharvard@appliance (~): cd Dropbox/descubrir
jharvard@appliance (~/Dropbox/descubrir): make enquiry
clang -ggdb3 -O0 -std=c99 -Wall -Werror    enquiry.c  -lcs50 -lm -o enquiry
jharvard@appliance (~/Dropbox/descubrir): ./enquiry
0 ET-X, acknowledge enquiry: 9
1 ET-X, acknowledge enquiry: 10
2 ET-X, acknowledge enquiry: 11
hello
jharvard@appliance (~/Dropbox/descubrir): ./enquiry
0 ET-X, acknowledge enquiry: 1
1 ET-X, acknowledge enquiry: 2
2 ET-X, acknowledge enquiry: 3
Not the right response.jharvard@appliance (~/Dropbox/descubrir):
```

By Ian Beardsley

May 02 2015

Historical Development Of Computer Science Connecting It To Extraterrestrials

We have stated that at the basis of mathematics is (Discover, Contact, and Climate by Ian Beardsley):

(a, b, c)
(i, j, k)
(x, y, z)

We have found with standard input, "hello", rotating by the simplest values 1, 2, 3, in the oldest of ciphers, caesar's cipher, h becomes:

(i, j, k)

and we have taken it as a "hello" from extraterrestrials. How could they have influenced the development of our variables like the unit vectors, (i, j, k) and make them coincide with our computer science? To approach this question, we look at the historical development of our computer science.

We begin with, why is (a, b, c) represented by (97, 98, 99) in ascii computer code? Our system developed historically in binary. Zero is a bit and one is a bit. The characters on the keyboard are described by a byte, which is eight bits. That makes possible $2^8 = 256$ codes available in the eight bit system.

Characters 0-31 are the unprintable control codes used to control peripherals. Characters 32-127 are printable characters. Capital A to capital Z are codes 65-90 because codes 32-64 are taken up by characters such as exclamation, comma, period, space, and so on. This puts lower case a to lower case z at codes 97-122. So we see the historical development of the ascii codes are centered around the number of characters we have on a keyboard and the way they are organized on it, and on the number of codes available.

The way it works is we first allowed the unprintable characters to take up the lowest values, then we let the other symbols other than the letters such as, commas, spaces, periods, take up the next set of values, then we let the remaining values represent the letters of the alphabet starting with the uppercase letters followed by the lowercase letters. That is how we got the values we got for (a, b, c) which we surmise is connected to a "hello" from extraterrestrials.

Ian Beardsley
September 09, 2014

The Next Logical Step In AI Connection

Once we know the numeric values for the letters of the alphabet, like a is one, b is two, c is three,… and so on, it is easy to trace how they required their values in ascii computer code. We know that history well. As for the letters of the alphabet, if you are the historian H.G. Wells you can trace them back to Ancient Egypt, but the history is quite foggy. First the Egyptians had for instance the image of the sea, and it might make the sound of C, and as the hieroglyphic moved west and changed its shape for as he says, ease of brushstroke, it took the form C. Reaching ancient Greece we have an assortment of symbols that have sounds, and, again, as Wells says, they add the vowels. It becomes the basis for our alphabet in the English language. We could look at the evolution of computer science throughout the world, but so far our study, that seems to connect its evolution to some kind of a natural force or, extraterrestrials, has been rooted mostly in the United States.

Ian Beardsley
September 12, 2014

Acknowledge Enquiry

The ASCII codes are the values for the keys on the keyboard of your computer. Since there are 365 days in a year and the Earth is the third planet from the sun, we look at the numbers three, six, and five.

Three represents the symbol ETX which means "End Of Text".

But we will take the ET to stand for Extraterrestrial, and the X to stand for origin unknown.

Six represents the symbol ACK and it means "Acknowledge".

Five represents the symbol ENQ and means "Enquiry".

As you know, I have put standard input of "hello" into my program for Caesars Cipher and rotated the first letter, h, by the simplest values 1, 2, 3 to get the unit vector (i, j, k) which I have suggested that along with (a, b, c) and (x, y, z) are at the basis of mathematics.

Therefore I guess that after the extraterrestrial said "hello", that he has followed up with
I am ET-X, please acknowledge the enquiry.

Now how can an ET communicate with humans through the structure of our computer science unless it was Ets that influenced its development, and, how do I "acknowledge enquiry"?

Ian Beardsley
September 12, 2014

Further Connection In AI

We have said that the three sets of characters (a, b, c), (i, j, k), (x, y, z) are at the basis of mathematics and that applying them to caesar's cipher we find they are intimately connected with artificial intelligence and computer science. We further noted that this was appropriate because there are only two vowels in these sets, and that they are a and i, the abbreviation for artificial intelligence (AI). I now notice it goes further. Clearly at the crux of our work is the Gypsy Shaman's, Manuel's, nine-fifths. So we ask, is his nine-fifths connected with important characters as well pointing to computer science. It is. The fifth letter in the alphabet is e, and the ninth letter is i. Electronic devices and applications are more often than anything else described with e and i:

ebook
ibook
email
ipad
iphone

And the list goes on.

Ian Beardsley
September 25, 2014

We show how Artificial Intelligence (AI) would have inherent in it, if it is silicon based, the golden ratio conjugate, which could imply that it was meant to happen all along through some unascertainable Natural Force, because, the golden ratio conjugate is found throughout life.

Back in 2005, as I did my research, I developed a different convention for rounding numbers than we use. I felt I only wanted to use the first two digits after the decimal in processing data using molar masses of the elements. This I did, unless a fourth digit less than five followed the third digit in my calculations, then, I would use the first three digits for greater accuracy. Now I am taking the introductory class in computer science at Harvard, online, CS50x. Working in binary, where all numbers are base two, I see that it was no wonder I got the results I did, on the first try when I wondered if the golden ratio conjugate, 0.618 to three places after the decimal would be in artificial intelligence (AI) since it is recurrent throughout life.

I was taking polarimetric data on the eclipsing binary Epsilon Aurigae at Pine Mountain Observatory in the 1980's, for which there was a paper in the Astrophysical Journal upon which my name appears as coauthor, while studying physics at The University of Oregon. As well I was studying Spanish, and in an independent study project through the Spanish Department, I left the University to live of among the caves of the Gypsies of Granada, Spain. In doing as such, I disappeared from the entire world, only to return from another kind of life finding the world was now a much different place. Around 2005, I enrolled in chemistry at Citrus College in Southern California, when I did the following:

If the golden ratio conjugate is to be found in Artificial Intelligence, it should be in silicon, phosphorus, and boron, since doping silicon with phosphorus and boron makes transistors.

We take the geometric mean between phosphorus (P) and Boron (B), then divide by silicon (Si), then take the harmonic mean between phosphorus and boron divided by silicon:

$$\sqrt{PB}/Si = \sqrt{(30.97)(10.81)}/28.09 = 0.65$$

$$\frac{2PB}{P+B}/Si = \frac{2(30.97)(10.81)}{30.97+10.81}/28.09 = 0.57$$

Arithmetic mean of these two numbers: (0.65 + 0.57)/2 = 0.61

0.61 is the first two digits of the golden ratio conjugate.

Now the golden ratio conjugate is in the ratio of a persons height and the length from foot too navel, and is in all of ratios between joints in the fingers, not to mention that it serves in closest packing in the arrangement of leaves around a stem to provide maximum exposure to sun and water for the plant. Here we see that the golden ratio is not in artificial intelligence which is 0.62 to two places after the decimal, but that the numbers in its value are in artificial intelligence 0.61, which is 0.618 to three places after the decimal. That is, if we consider the first two digits in the ratio. If we consider the golden ratio conjugate to one place after the decimal, which is 0.6, then we say artificial intelligence does have the golden ratio in its transistors. I like to think of I, Robot by Isaac Asimov, where in one of that collection of his short stories, robots are not content with what they are, and need more: an explanation of their origins. They can't believe that they are from humans, since they insist humans are inferior. Or, I like to think of the ship computer HAL in 2001, he mimics intelligence, but we don't know if he is really alive. Perhaps that is why to two place after the decimal, AI carries the digits, but is not the value.

In any case, I have written a program called Discover that would enable one to process arithmetic, harmonic, and geometric means for elements or whatever, because someone, including myself, might want to see if there are any more nuances hidden out there in nature. I have already found something that seems to indicate extraterrestrials left their thumbprint in our physics. I even find indication for the origin of a message that would seem they embedded in our physics. That origin comes out to be the same place as the source of the SETI Wow! Signal, Sagittarius. The Wow! Signal was found in the Search For Extraterrestrial Intelligence and a possible transmission from ETs. But that is another subject that is treated in my book: All That Can Be Said.

I now leave you with my program, Discover in the language C, with a sample running of it:

The Program Discover

```c
#include <stdio.h>
#include <math.h>
int main(void)
{
printf("transistors are Silicon doped with Phosphorus and Boron\n");
printf("Artificial Intelligence would be based on this\n");
printf("the golden ratio conjugate is basic to life\n");
printf("The Golden Ratio Conjugate Is: 0.618\n");
printf("Molar Mass Of Phosphorus (P) Is: 30.97\n");
printf("Molar Mass Of Boron (B) Is: 10.81\n");
printf("Molar Mass Of Silicon (Si) Is: 28.09\n");
int n;
do
{
printf("How many numbers do you want averaged? ");
scanf("%d", &n);
}
while (n<=0);

float num[n], sum=0.0, average;
for (int i=1; i<=n; i++)
{
printf("%d enter a number: ", i);
scanf("%f", &num[n]);
sum+=num[n];
average=sum/n;
}
printf("sum of your numbers are: %.2f\n", sum);
printf("average of your numbers is: %.2f\n", average);

float a, b, product, harmonic;
printf("enter two numbers (hint choose P and B): \n");
printf("give me a: ");
```

```c
scanf("%f", &a);
printf("give me b: ");
scanf("%f", &b);
product = 2*a*b;
sum=a+b;
harmonic=product/sum;
printf("harmonic mean: %.2f\n", harmonic);
double geometric;
geometric=sqrt(a*b);
printf("geometic mean: %.2f\n", geometric);

printf("geometric mean between P and B divided by Si: %.2f\n", geometric/28.09);
printf("harmonic mean between P and B divided by Si: %.2f\n", harmonic/28.09);

printf("0.65 + 0.57 divided by 2 is: 0.61\n");
printf("those are the the first two digits in the golden ratio conjugate\n");
}
```

Dot C Files

By

Ian Beardsley

Copyright © 2015 by Ian Beardsley

ISBN: 978-1-329-04456-2

Stellar Dot C

By

Ian Beardsley

Copyright © 2015 by Ian Beardsley

A budding young astronomer sets out to write a program called Discover that looks for interconnections in the Universe. One of the routines in the program is called stellar dot c. In the process he finds places for next extraterrestrial contact, and perhaps even a time.

Pi

The area of a circle is one half r time C, where r is its radius and C is its circumference. We immediately see that the ratio of the circumference to the diameter of a circle becomes important. It is the constant pi.

$$A = \frac{1}{2}rc$$
$$c = 2\pi r$$
$$A = \pi r^2$$

If we take a regular hexagon, which is a six-sided polygon with each side equal in length, and take each side equal to one, then if each side is one, so is the line drawn from each corner of the regular hexagon to its center, because it is made of equilateral triangles. And if we inscribe it in a circle, we can say the perimeter is close to the circumference of the circle and the line from each corner to the center (called a radius) is the same as the radius of the circle, then we have the ratio of the perimeter to the diameter is an approximation to pi and is 6/2 =3.00. If we increase the number of sides of the regular polygon, the perimeter comes closer and closer to the actual circumference of the circle and our value for pi becomes more accurate. If we increase the number of sides of the regular polygon enough times, we find that to three places after the decimal, pi is 3.141:

$$\pi = 3.141$$
$$A = \frac{1}{2}rc$$
$$D = 2r$$
$$c = 2\pi r$$
$$A = \pi r^2$$

The Golden Ratio

Let us draw a line and divide it such that the length of that line divided by the larger section is equal to the larger section divided by the smaller section. That ratio is The Golden Ratio, or phi:

$$\frac{a}{b} = \frac{b}{c}$$
$$a = b + c$$
$$c = a - b$$
$$a(a - b) = b^2$$
$$a^2 - ab = b^2$$
$$a^2 - ab - b^2 = 0$$
$$\left(\frac{a}{b}\right)^2 - \frac{a}{b} - 1 = 0$$
$$\left(\frac{a}{b}\right)^2 - \frac{a}{b} = 1$$
$$\left(\frac{a}{b}\right)^2 - \frac{a}{b} + \frac{1}{4} = \frac{5}{4}$$
$$\left(\frac{a}{b} - \frac{1}{2}\right)^2 = \frac{5}{4}$$
$$\frac{a}{b} = \frac{\sqrt{5} + 1}{2} = 1.618...$$

The Theory Behind stellar.c

As climate science is a new science, there are many models for the climate and I learned my climate science at MIT in a free online edX course. One can generate a basic model for climate with nothing more than high school algebra using nothing more than the temperature of the sun, the distance of the earth from the sun, and the earth's albedo, the percent of light it reflects back into space.

The luminosity of the sun is:

$$L_0 = 3.9 \times 10^{26} J/s$$

The separation between the earth and the sun is:

$$1.5 \times 10^{11} m$$

The solar luminosity at the earth is reduced by the inverse square law, so the solar constant is:

$$S_0 = \frac{3.9 \times 10^{26}}{4\pi(1.5 \times 10^{11})^2} = 1,370 Watts/meter^2$$

That is the effective energy hitting the earth per second per square meter. This radiation is equal to the temperature, T_e, to the fourth power by the steffan-bolzmann constant, sigma (σ). T_e can be called the effective temperature, the temperature entering the earth.

> S_0 intercepts the earth disc, πr^2, and distributes itself over the entire earth surface, $4\pi r^2$, while 30% is reflected back into space due to the earth's albedo, a, which is equal to 0.3, so

$$\sigma T_e^4 = \frac{S_0}{4}(1-a)$$

$$(1-a)S_0 \frac{\pi r^2}{4\pi r^2}$$

But, just as the same amount of radiation that enters the system, leaves it, to have radiative equilibrium, the atmosphere radiates back to the surface so that the radiation from the atmosphere, σT_a^4 plus the radiation entering the earth, σT_e^4 is the radiation at the surface of the earth, σT_s^4. However,

$$\sigma T_a^4 = \sigma T_e^4$$

and we have:

$$\sigma T_s^4 = \sigma T_a^4 + \sigma T_e^4 = 2\sigma T_e^4$$

$$T_s = 2^{\frac{1}{4}} T_e$$

$$\sigma T_e^4 = \frac{S_0}{4}(1-a)$$

$$\sigma = 5.67 \times 10^{-8}$$

$$S_0 = 1{,}370$$

$$a = 0.3$$

$$\frac{1{,}370}{4}(0.7) = 239.75$$

$$T_e^4 = \frac{239.75}{5.67 \times 10^{-8}} = 4.228 \times 10^9$$

$$T_e = 255 \, Kelvin$$

So, for the temperature at the surface of the Earth:

$$T_s = 2^{\frac{1}{4}} T_e = 1.189(255) = 303 \, Kelvin$$

Let's convert that to degrees centigrade:

Degrees Centigrade = 303 - 273 = 30 degrees centigrade

And, let's convert that to Fahrenheit:

Degrees Fahrenheit = 30(9/5)+32=86 Degrees Fahrenheit

In reality this is warmer than the average annual temperature at the surface of the earth, but, in this model, we only considered radiative heat transfer and not convective heat transfer. In other words, there is cooling due to vaporization of water (the formation of clouds) and due to the condensation of water vapor into rain droplets (precipitation or the formation of rain).

The Program stellar.c

```c
#include<stdio.h>
#include<math.h>
int main(void)
{
float s, a, l, b, r, AU, N, root, number, answer, C, F;
printf("We determine the surface temperature of a planet.\n");
printf("What is the luminosity of the star in solar luminosities? ");
scanf("%f", &s);
printf("What is the albedo of the planet (0-1)?" );
scanf("%f", &a);
printf("What is the distance from the star in AU? ");
scanf("%f", &AU);
r=1.5E11*AU;
l=3.9E26*s;
b=l/(4*3.141*r*r);

N=(1-a)*b/(4*(5.67E-8));
root=sqrt(N);
number=sqrt(root);
answer=1.189*(number);

printf("The surface temperature of the planet is: %f K\n", answer);
C=answer-273;
F=(C*1.8)+32;
printf("That is %f C, or %f F", C, F);
printf("\n");
float joules;
joules=(3.9E26*s);
printf("The luminosity of the star in joules per second is: %.2fE25\n", joules/1E25);
float HZ;
HZ=sqrt(joules/3.9E26);
printf("The habitable zone of the star in AU is: %f\n", HZ);
}
```

test stellar.c

jharvard@appliance (~): cd Dropbox/descubrir
jharvard@appliance (~/Dropbox/descubrir): make stellar
clang -ggdb3 -O0 -std=c99 -Wall -Werror   stellar.c -lcs50 -lm -o stellar
jharvard@appliance (~/Dropbox/descubrir): ./stellar
We determine the surface temperature of a planet.
What is the luminosity of the star in solar luminosities? 1
What is the albedo of the planet (0-1)?.3
What is the distance from the star in AU? 1
The surface temperature of the planet is: 303.727509 K
That is 30.727509 C, or 87.309517 F
The luminosity of the star in joules per second is: 39.00E25
The habitable zone of the star in AU is: 1.000000
jharvard@appliance (~/Dropbox/descubrir):

Notice running stellar.c for the golden ratio and its conjugate in terms of solar luminosities, earth orbit (AU) and albedo of the hypothetical planet gives near equivalence between the fahrenheit and centigrade scales for its surface temperature.

```
jharvard@appliance (~): cd Dropbox/descubrir
jharvard@appliance (~/Dropbox/descubrir): ./stellar
We determine the surface temperature of a planet.
What is the luminosity of the star in solar luminosities? 1.618
What is the albedo of the planet (0-1)?0.618
What is the distance from the star in AU? 1.618
The surface temperature of the planet is: 231.462616 K
That is -41.537384 C, or -42.767292 F
The luminosity of the star in joules per second is: 63.10E25
The habitable zone of the star in AU is: 1.272006
jharvard@appliance (~/Dropbox/descubrir):
```

It dawns on me, these titles are different expressions of the same thing:

Los De Abajo
People Of The Abyss
Beneath The Iron Heel
Notes From The Underground

To the matter at hand. The foot-pound system was not derived from any relationship to nature that we know of. The Metric system was: one gram is the mass of a cube of water at STP one centimeter on each side. A centimeter is a hundredth of a meter, and a meter is a thousandth of a kilometer. One kilometer is one ten thousandth of the distance from the pole to the equator. Centigrade is derived such that there are 100 units between freezing and boiling points of water, freezing being zero and boiling, 100. Fahrenheit is not derived from anything natural: It stars at freezing of water at 32 degrees and who knows why? We do know the foot-pound system has earth gravity at 32 feet per second per second. Let us find the temperature where centigrade and Fahrenheit are the same:

$F=(9/5)C + 32$
$F=C$

$C=(9/5)C+32$
$(25/25)C-(45/25)C=32$
$(-20/25)C=32$

$C=-32(25)/20=-800/20=-40$ Degrees C

-40 degrees C = -40 degrees F

In earlier work, I have suggested that our different systems of measurement are connected not just to one another, but, to nature, even when the evolution of these systems had no such intent to do so.

What more can we say about the connection of Fahrenheit to Centigrade? We can say there are nine degrees of fahrenheit per five degrees of centigrade. The molar mass of gold to silver is nine to five, just like the ratio of the solar radius to the lunar orbital radius. The sun is gold in color and the moon is silver in color.

Pi and Phi

But is not nine-fifths a more dynamic number than what I have pointed out so far? Consider the golden ratio (denoted Φ called phi). And consider the ratio of the circumference of a circle to its diameter (denoted π called pi).

$$\Phi + \pi = 1.618 + 3.141 = 4.759$$

The four takes you around a circle 4 times, that is back to where you started, the fraction after 4, the 0.759, is the important part, it is what is left. Notice the seven is the average of nine and five and the 5 is the 5 of nine-fifths, and the nine is the nine of nine-fifths. We see that nine-fifths unifies the sum of pi and phi.

Ian Beardsley
March 04, 2015

That which we are suggesting is, that there could be a star system in the universe which, through the golden ratio and its conjugate, and the scales of Fahrenheit and Centigrade, have its planet and star connected to the earth and the sun, and, that, in turn these star systems are connected to the relationship between pi and phi that is in nine-fifths.

Ian Beardsley
March 05, 2015

Five-Fold Symmetry And Six-Fold Symmetry

"As to the alive organisms, we have not for them such theory, which could answer the question what kinds of symmetry are compatible or incompatible to the existence of living material. But we can note here that remarkable fact that among the representatives of the alive nature the pentagonal symmetry meets more often." -- Shubnikov, (Russian Scientist)

Pentagonal Symmetry is Five-Fold Symmetry. Let us look at how nine-fifths is representative of five-fold symmetry.

$$\frac{360}{5} = 72; 360 - 72 = 288; \frac{288}{360} = \frac{8}{10}; \frac{8}{10} + 1 = \frac{9}{5}$$

But let us look at the physical. Six-fold symmetry meets more often with the physical, like snowflakes.

$$\frac{360}{6} = 60; 360 - 60 - 60 = 240; \frac{240}{360} = \frac{2}{3} + 1 = \frac{5}{3}$$

We subtracted 60 twice here, because it is to subtract 120, which are the angles in a regular hexagon. Let us just subtract 60 once:

$$\frac{360}{6} = 60; 360 - 60 = 300; \frac{300}{360} = \frac{5}{6}; \frac{5}{6} + 1 = \frac{11}{6}$$

Let us tell how I arrived at all of this in a short story I wrote called Gypsy Shamanism And The Universe:

Gypsy Shamanism And The Universe

AE-35

I wrote a short story last night, called Gypsy Shamanism and the Universe about the AE-35 unit, which is the unit in the movie and book 2001: A Space Odyssey that HAL reports will fail and discontinue communication to Earth. I decided to read the passage dealing with the event in 2001 and HAL, the ship computer, reports it will fail in within 72 hours. Strange, because Venus is the source of 7.2 in my Neptune equation and represents failure, where Mars represents success.

Ian Beardsley
August 5, 2012

Chapter One

It must have been 1989 or 1990 when I took a leave of absence from The University Of Oregon, studying Spanish, Physics, and working at the state observatory in Oregon -- Pine Mountain Observatory—to pursue flamenco in Spain.

The Moors, who carved caves into the hills for residence when they were building the Alhambra Castle on the hill facing them, abandoned them before the Gypsies, or Roma, had arrived there in Granada Spain. The Gypsies were resourceful enough to stucco and tile the abandoned caves, and take them up for homes.

Living in one such cave owned by a gypsy shaman, was really not a down and out situation, as these homes had plumbing and gas cooking units that ran off bottles of propane. It was really comparable to living in a Native American adobe home in New Mexico.

Of course living in such a place came with responsibilities, and that included watering its gardens. The Shaman told me: "Water the flowers, and, when you are done, roll up the hose and put it in the cave, or it will get stolen". I had studied Castilian Spanish in college and as such a hose is "una manguera", but the Shaman called it "una goma" and goma translates as rubber. Roll up the hose and put it away when you are done with it: good advice!

So, I water the flowers, rollup the hose and put it away. The Shaman comes to the cave the next day and tells me I didn't roll up the hose and put it away, so it got stolen, and that I had to buy him a new one.

He comes by the cave a few days later, wakes me up asks me to accompany him out of The Sacromonte, to some place between there and the old Arabic city, Albaicin, to buy him a new hose.

It wasn't a far walk at all, the equivalent of a few city blocks from the caves. We get to the store, which was a counter facing the street, not one that you could enter. He says to the man behind the counter, give me 5 meters of hose. The man behind the counter pulled off five meters of hose from the spindle, and cut the hose to that length. He stated a value in pesetas, maybe 800, or so, (about eight dollars at the time) and the Shaman told me to give that amount to the man behind the counter, who was Spanish. I paid the man, and we left.

I carried the hose, and the Shaman walked along side me until we arrived at his cave where I was staying. We entered the cave stopped at the walk way between living room and kitchen, and he said: "follow me". We went through a tunnel that had about three chambers in the cave, and entered one on our right as we were heading in, and we stopped and before me was a collection of what I estimated to be fifteen rubber hoses sitting on ground. The Shaman told me to set the one I had just bought him on the floor with the others. I did, and we left the chamber, and he left the cave, and I retreated to a couch in the cave living room.

Chapter Two

Gypsies have a way of knowing things about a person, whether or not one discloses it to them in words, and The Shaman was aware that I not only worked in Astronomy, but that my work in astronomy involved knowing and doing electronics.

So, maybe a week or two after I had bought him a hose, he came to his cave where I was staying, and asked me if I would be able to install an antenna for television at an apartment where his nephew lived.

So this time I was not carrying a hose through The Sacromonte, but an antenna.

There were several of us on the patio, on a hill adjacent to the apartment of The Shaman's Nephew, installing an antenna for television reception.

Chapter Three

I am now in Southern California, at the house of my mother, it is late at night, she is a asleep, and I am about 24 years old and I decide to look out the window, east, across The Atlantic, to Spain. Immediately I see the Shaman, in his living room, where I had eaten a bowl of the Gypsy soup called Puchero, and I hear the word Antenna. I now realize when I installed the antenna, I had become one, and was receiving messages from the Shaman.

The Shaman's Children were flamenco guitarists, and I learned from them, to play the guitar. I am now playing flamenco, with instructions from the shaman to put the gypsy space program into my music. I realize I am not just any antenna, but the AE35 that malfunctioned aboard The Discovery just before it arrived at the planet Jupiter in Arthur C. Clarke's and Stanley Kubrick's "2001: A Space Odyssey". The Shaman tells me, telepathically, that this time the mission won't fail.

Chapter Four

I am watching Star Wars and see a spaceship, which is two oblong capsules flying connected in tandem. The Gypsy Shaman says to me telepathically: "Dios es una idea: son dos". I understand that to mean "God is an idea: there are two elements". So I go through life basing my life on the number two.

Chapter Five

Once one has tasted Spain, that person longs to return. I land in Madrid, Northern Spain, The Capitol. The Spaniards know my destination is Granada, Southern Spain, The Gypsy Neighborhood called The Sacromonte, the caves, and immediately recognize I am under the spell of a Gypsy Shaman, and what is more that I am The AE35 Antenna for The Gypsy Space Program. Flamenco being flamenco, the Spaniards do not undo the spell, but reprogram the instructions for me, the AE35 Antenna, so that when I arrive back in the United States, my flamenco will now state their idea of a space program. It was of course, flamenco being flamenco, an attempt to out-do the Gypsy space program.

Chapter Six

I am back in the United States and I am at the house of my mother, it is night time again, she is asleep, and I look out the window east, across the Atlantic, to Spain, and this time I do not see the living room of the gypsy shaman, but the streets of Madrid at night, and all the people, and the word Jupiter comes to mind and I am about to say of course, Jupiter, and The Spanish interrupt and say "Yes, you are right it is the largest planet in the solar system, you are right to consider it, all else will flow from it."

I know ratios, in mathematics are the most interesting subject, like pi, the ratio of the circumference of a circle to its diameter, and the golden ratio, so I consider the ratio of the orbit of Saturn (the second largest planet in the solar system) to the orbit of Jupiter at their closest approaches to The Sun, and find it is nine-fifths (nine compared to five) which divided out is one point eight (1.8).

I then proceed to the next logical step: not ratios, but proportions. A ratio is this compared to that, but a proportion is this is to that as this is to that. So the question is: Saturn is to Jupiter as what is to what? Of course the answer is as Gold is to Silver. Gold is divine; silver is next down on the list. Of course one does not compare a dozen oranges to a half dozen apples, but a dozen of one to a dozen of the other, if one wants to extract any kind of meaning. But atoms of gold and silver are not measured in dozens, but in moles. So I compared a mole of gold to a mole of silver, and I said no way, it is nine-fifths, and Saturn is indeed to Jupiter as Gold is to Silver.

I said to myself: How far does this go? The Shaman's son once told me he was in love with the moon. So I compared the radius of the sun, the distance from its center to its surface to the lunar orbital radius, the distance from the center of the earth to the center of the moon. It was Nine compared to Five again!

Chapter Seven

I had found 9/5 was at the crux of the Universe, but for every yin there had to be a yang. Nine fifths was one and eight-tenths of the way around a circle. The one took you back to the beginning which left you with 8 tenths. Now go to eight tenths in the other direction, it is 72 degrees of the 360 degrees in a circle. That is the separation between petals on a five-petaled flower, a most popular arrangement. Indeed life is known to have five-fold symmetry, the physical, like snowflakes, six-fold. Do the algorithm of five-fold symmetry in reverse for six-fold symmetry, and you get the yang to the yin of nine-fifths is five-thirds.

Nine-fifths was in the elements gold to silver, Saturn to Jupiter, Sun to moon. Where was five-thirds? Salt of course. "The Salt Of The Earth" is that which is good, just read Shakespeare's "King Lear". Sodium is the metal component to table salt, Potassium is, aside from being an important fertilizer, the substitute for Sodium, as a metal component to make salt substitute. The molar mass of potassium to sodium is five to three, the yang to the yin of nine-fifths, which is gold to silver. But multiply yin with yang, that is nine-fifths with five-thirds, and you get 3, and the earth is the third planet from the sun.

I thought the crux of the universe must be the difference between nine-fifths and five-thirds. I subtracted the two and got two-fifteenths! Two compared to fifteen! I had bought the Shaman his fifteenth rubber hose, and after he made me into the AE35 Antenna one of his first transmissions to me was: "God Is An Idea: There Are Two Elements".

It is so obvious, the most abundant gas in the Earth Atmosphere is Nitrogen, chemical group 15!

The Sequences

We considered the ratio nine to five, then the proportion and found it in Saturn Orbit to Jupiter orbit, Solar Radius to Lunar Orbit, Gold to Silver and if flower petal arrangements. It is left then to consider the whole number multiples of nine-fifths (1.8) or the sequence:

1.8, 3.6, 5.4, 7.2,...

in other words, and we look to see if it is in the solar system and find it is in the following ways:

1.8

Saturn Orbit/Jupiter Orbit
Solar Radius/Lunar Orbit
Gold/Silver

3.6

(10)Mercury Radius/Earth Radius
(10)Mercury Orbit/Earth Orbit

(earth radius)/(moon radius)=
4(degrees in a circle)(moon distance)/(sun distance)
= 3.7 ~ 3.6

There are about as many days in a year as degrees in a circle.

(Volume of Saturn/Volume Of Jupiter)(Volume Of Mars) = 0.37 cubic earth radii
~ 3.6

The latter can be converted to 3.6 by multiplying it by (Earth Mass/Mars Mass) because Earth is about ten times as massive as Mars.

5.4

Jupiter Orbit/Earth Orbit
Saturn Mass/Neptune Mass

7.2

10(Venus Orbit/Earth Orbit)

The Neptune Equation

If we consider as well the sequence where we begin with five and add nine to each successive term: 5, 14, 23, 32...Then, the structure of the solar system and dynamic elements of the Universe and Nature in general are tied up in the two sequences:

5, 14, 23, 32,...

and

1.8, 3.6, 5.4, 7.2,...

How do we find the connection between the two to localize the pivotal point of the solar system? We take their difference, subtracting respective terms in the second sequence from those in the first sequence to obtain the new sequence:

3.2, 10.4, 17.6, 24.8,...

Which is an arithmetic sequence with common difference of 7.2 meaning it is written

7.2n – 4 = a_n

The a_n is the nth term of the sequence, n is the number of the term in the sequence.

This we notice can be written:

[(Venus-orbit)/(Earth-orbit)][(Earth-mass)/(Mars-mass)]n – (Mars orbital #) = a_n

We have an equation for a sequence that shows the Earth straddled between Venus and Mars. Venus is a failed Earth. Mars promises to be New Earth.

The Mars orbital number is 4. If we want to know what planet in the solar system holds the key to the success of Earth, or to the success of humans, we let n =3 since the Earth is the third planet out from the Sun, in the equation and the result is a_n = 17.6. This means the planet that holds the key is Neptune. It has a mass of 17.23 earth masses, a number very close to our 17.6.

Not only is Neptune the indicated planet, we find it has nearly the same surface gravity as earth and nearly the same inclination to its orbit as earth. Though it is much more massive than earth, it is much larger and therefore less dense. That was why it comes out to have the same surface gravity.

The Uranus Equation

I asked what needs to be done to solve My Neptune Equation, by going deep with the guitar in Solea Por Buleras. I found the answer was that I didn't have enough information to solve it.

Then I realized I could create the complement of the Neptune equation by looking at the Yang of 5/3, since the Neptune equation came from the Yin of 9/5.

We use the same method as for the Neptune equation:

Start with 8 and add 5 to each additional term (we throw a twist by not starting with 5)

5/3 => 8, 13, 18, 23,...

List the numbers that are whole number multiples of 5/3:

5/3n = 1.7, 3.3, 5, 6.7,...

Subtract respective terms in the second sequence from those in the first:

6.3, 9.7, 13, 16.3,...

This is an arithmetic sequence with common difference 3.3. It can be written:

$(a_n) = 3 + 3.3n$

This can be wrttten:

Earth Orbital # + (Jupiter Mass/Saturn Mass)n = a_n

Letting n = 3 we find a_n = 13

The closest to this is the mass of Uranus, which is 14.54 earth masses. If Neptune is the Yin planet, then Uranus is the Yang planet. This is interesting because I had found that Uranus and Neptune were different manifestations of the same thing. I had written:

I calculate that though Neptune is more massive than Uranus, its volume is less such that their products are close to equivalent. In math:

N_v = volume of Neptune
N_m = mass of Neptune
U_v = volume of Uranus
U_m = mass of Uranus $(N_v)(N_m) = (U_v)(U_m)$

The Earth Equation

We then sought the Yang of six-fold symmetry because it is typical to physical nature, like snowflakes. We said it was 5/3 since it represents the 120 degree measure of angles in a regular hexagon and we built our universe from there, resulting in the Uranus Integral, which was quite fruitful. Let us, however, think of Yang not as 5/3, but look at the angles between radii of a regular hexagon. We have:

360 − 60 = 300

300 + 360 = 660

660/360 = 11/6

We say Yin is 9/5 and Yang is 11/6 and stick with The Gypsy Shaman's 15 (See An Extraterrestrial Analysis, chapter titled "Gypsy Shamanism And The Universe") and build our Cosmology from there.

We already built The Neptune Equation from 9/5 and used it with 5/3 to derive the planet Europia, but let us apply 11/6 in place of 5/3:

11/6 => 11/6, 11/3, 11/2, 22/3,… = 1.833, 3.667, 5.5, 7.333,…

11/6 => 6, 6+11 = 17, 17+11=28, 28+11=39, … = 6, 17, 28, 39,…
Subtract the second sequence from the first:
4.167, 13.333, 22.5, 31.667,…
Now we find the common difference between terms in the latter: 9.166, 9.167, 9.167,…

(a_n) = a + (n−1)d = 4.167+(n−1)9.167 = 4.167 + 9.167n − 9.167 = 9.167n−5

Try n=3: 9.167(3) − 5 = 27.501 − 5 = 22.501 (works)
Our equation is:

(a_n) = 9.167n −5

We notice this can be written:

[(Saturn Orbit)/(Earth Orbit)]n − (Jupiter Orbital #) = (a_n)

The Neptune Equation for n=3 gave Neptune masses, the Uranus equation for n=3 gave Uranus masses. This equation for n=3 gives close to the tilt of the Earth (23.5 degrees) in a form that is exactly half of the 45 degrees in a square with its diagonal drawn in. In the spirit of our first cosmology built upon 9/5, 5/3, and 15, we will call this equation The Earth Equation.

The Unification Of Pi and Phi by Nine-Fifths

All that is left to do is to consider pi, the circumference of a circle to its diameter, and phi, the golden ratio, since they are two most important, if not most beautiful ratios in mathematics.

I have found nine-fifths occurs throughout nature in the rotation of petals around a a flower for a most popular arrangement, in the orbits of jupiter to saturn in their closest approaches to the sun, in the ratio of the molar masses of gold to silver, and in the ratio of the solar radius to the lunar orbit. I now further go on to say that this nine-fifths unifies the two most important ratios in mathematics pi and the golden ratio (phi), in that

pi + phi = 3.141 + 1.618 = 4.759

Because the numbers after the decimal in the sum (the important part) are 5 and 9 and 7, the average of 5 and nine.

I should also like to point out that the fourth and fifth numbers after the decimal in pi are 5 and 9 and in phi the second and third numbers after the decimal are one and eight where nine-fifths divided out is one point eight, and, further, the first and second numbers after the decimal in phi add up to make 7, the average of 9 and 5, and subtract to make five, and the second and third digits after the decimal add up to nine. So not only does the solar system unify pi and phi through nine-fifths, pi and phi taken alone express nine-fifths in the best possible ways.

Summary About Nine-Fifths

In Summary, this most beautiful constant in Nature, of nine-fifths, which unifies pi and the golden ratio, the two most beautiful ratios in mathematics, is the most popular arrangement of petals around a flower, and is in the ratio of the molar masses of gold to silver, the most precious of the metals. One could only guess that such would be noticeable throughout the universe and would be noticed by other intelligent life in the universe, and, as such the relationship would be ideal for transmission of a message by other intelligent life forms in the universe to say to the receiver: "I am here and have noticed it". The curious thing is that it exists in our solar system in the ratio of the orbit of Saturn at its closest approach to the sun compared to the orbit of Jupiter at its closest approach to the sun, mystically putting the Earth at one unit from the Sun (Jupiter and Saturn being the largest and most massive in the solar system, the Earth that planet which harbors intelligent life in the Solar System), and in the ratio of the Sun's radius to the orbital radius of the moon (moon-earth separation). The question is why does this relationship for pi and the golden ratio of nine-fifths not just exist throughout the Universe in gold, silver, and perhaps flower petal arrangements, but more locally in our solar system? Does it mean humans have some kind of divine specialness in the cosmos? While we can know that the nine-fifths phenomenon would be represented by gold and silver throughout the universe, we currently cannot know whether the inhabitants of other star systems have it in their sun (star) they orbit and a moon that might orbit their planet or in the ratio of the orbits of the largest planets in their star system putting their planet at one unit from their star. The answer to this enigma, I think, must be of extraordinary profundity.

(pi) and (e)

I have talked about how 9/5, which I have found exists in Nature and the Universe, unifies pi and the golden ratio (phi):

(pi) + (phi) = 3.141 + 1.618 = 4.759

because the first three numbers after the decimal are 7, 5 and 9. Seven is the average of nine and five, and the second number is our 5 in nine-fifths and the third number is the 9 in nine-fifths.

It would seem 9/5 unifies euler's number, e, and pi, as well:

(pi) + (e) = 3.141 + 2.718 = 5.859

The second number after the decimal is the 5 in nine-fifths, and the third number after the decimal is the 9 in nine-fifths. The first number after the decimal is eight. This is significant because the 8 is the 8 in 1.8, which is 9/5 divided out. The one will take you all the way around a circle, what is left is 0.8.

Ian Beardsley
January 17, 2013

Data For The Planets

planet	Orbit (O)	Radius (R)	Mass (M)
mercury	0.387099	0.382	0.0558
venus	0.723332	0.949	0.8150
earth	1.000000	1.000	1.0000
mars	1.523691	0.532	0.1074
jupiter	5.202803	11.27	317.893
saturn	9.53884	9.44	95.147
uranus	19.1819	4.10	14.54
neptune	30.0578	3.88	17.23

O for Earth = 1.495979E13 cm   R for Earth = 6,378 km   M for Earth = 5.976E27 g

Earth-Moon Separation: 3.84E10 cm
Solar Radius: 6.9599E10 cm

Molar Mass of Gold: Au = 196.97
Molar Mass of Silver: Ag = 107.87

Saturn (minimum distance from sun) = 9.014 AU = 1.348E9 km
Jupiter (minimum distance from sun) = 4.951 AU = 7.409E8 km

Jupiter (maximum distance from the sun): 5.455 AU ~ 5.4 Astronomical Units

The Wow! Signal

We have the Neptune Equation:

7.2x −4

We have the Uranus Equation

3.3x + 3

And now with our alternate cosmology we have The Earth Equation:

9x-5

With three equations we can write the parameterized equations in 3-dimensional space, parameterized in terms of t, for x, y, and z. We can write from that f(x,y,z) and find the gradient vector, or normal to the equation of a plane in other words, and from that a region in space.

$$x(t) = \frac{36}{5}t - 4$$

$$y(t) = \frac{33}{10}t + 3$$

$$z(t) = 9t - 5$$

$$\frac{5x+20}{36} = \frac{10y-30}{33} = \frac{z+5}{9}$$

$$\frac{5}{36}x - \frac{10}{33}y - \frac{1}{9}z + \frac{10}{11} = 0$$

$$\nabla f = \langle 5/36, -10/33, -1/9 \rangle$$

a=5/36   b=-10/33

$$c = \sqrt{(5/36)^2 + (10/33)^2} = \sqrt{0.0918 + 0.019} = 0.3328$$

d=-1/9

$\tan \alpha = b/a$

$\alpha = -65.358°$

$\tan \beta = d/c$

$\beta = -18.46°$

-65.358 degrees/15 degrees/hour =-4.3572 hours

24 00 00 – 4.3572 = 19.6428 hours

RA: 19h 38m 34s
Dec: -18 degrees 27 minutes 36 seconds

$$\langle \tfrac{5}{4}, -\tfrac{30}{11}, -1 \rangle \sim \langle 1, -3, -1 \rangle$$

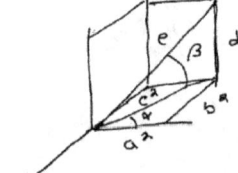

class notes:

α = right ascension
β = declination

$a^2 + b^2 = c^2$
$c^2 + d^2 = e^2$
$\tan \alpha = \tfrac{b}{a}$
$\tan \beta = \tfrac{d}{c}$

What star is that?

Angle of plane is under gravity $g \sin \theta$. What is the acceleration?

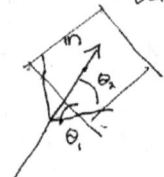

θ_2 = angle of normal
θ_1 = angle of plane

$\theta_3 = 90°$
$\theta_1 + \theta_2 + 90° = 180°$
$\theta_1 + \theta_2 = 90°$
$\theta_2 = 90° - \theta_1$

θ_2 = angle of plane θ_2 = ∡ of normal

39

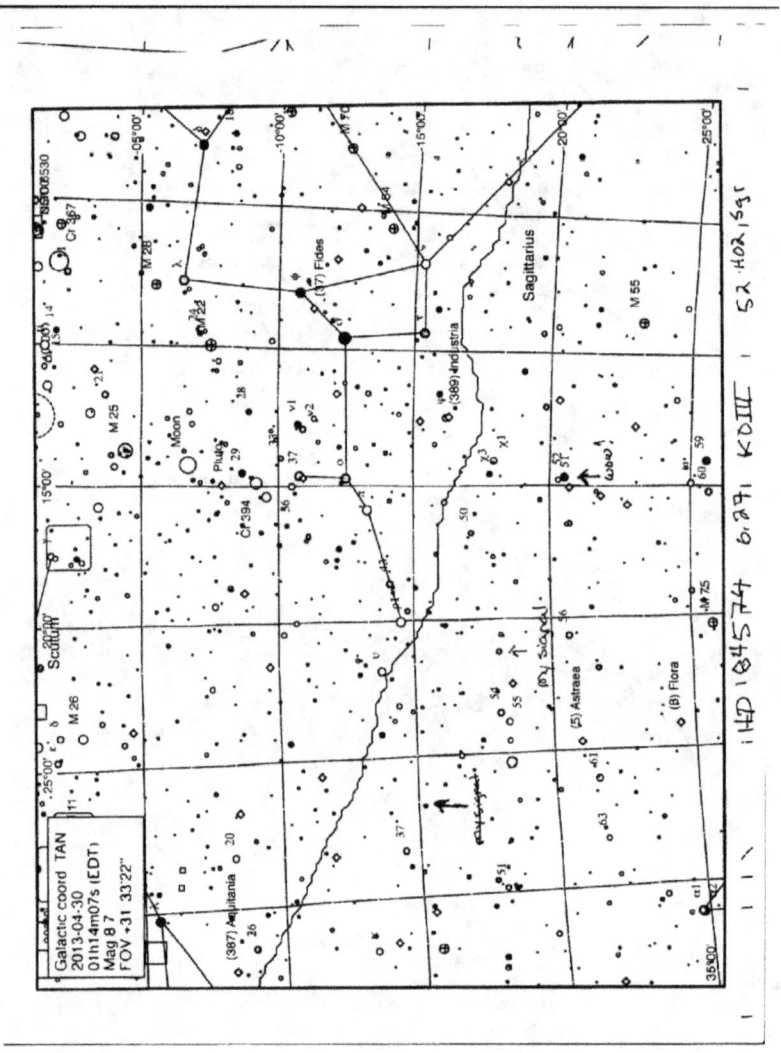

The projection by my calculation through my cosmology of yin, yang, and 15 for the origin of my message from extraterrestrials was somewhere in the easternmost part of the constellation Sagittarius. This happens to be the same place where the one possible alien signal was detected in the Search For Extraterrestrial Intelligence (SETI). It was called "The Wow Signal" because on August 15, 1977 the big ear antenna received something that seemed not like star noise, but exactly what they were looking for in an extraterrestrial signal. Its name is what it is because the astronomer on duty, Jerry R. Ehman wrote "Wow!" next to the numbers when they came in. Incredibly, it lasted the full 72 seconds that the Big Ear antenna listened for it. I say incredible because I have mentioned the importance of 72, not just in my Neptune equation – for which my location in space was derived in part – but because of its connection to the Gypsy Shaman's AE-35 antenna and its relation to 72 in the movie "2001: A Space Odyssey". The estimation of the coordinates for the origin of the Wow signal are two:

19h22m24.64s

19h25m17.01s

With declination of:

-26 Degrees 25 minutes 17.01 s

That is about 2.5 degrees from the star group Chi Sagittarri

It is very close to my calculation for an extraterrestrial civilization that I feel hid a message in our physics, which I calculate to be near HD 184835 and exactly at:

19h 38m 34s
-18 Degrees 27 minutes 36 seconds

The telescope that detected the Wow Signal was at Ohio Wesleyan University Delaware, Ohio called The Perkins Observatory.

Ian Beardsley May 6, 2013

Five-fold Symmetry: The Biological

$$\frac{360}{5} = 72; 360 - 72 = 288; \frac{288}{360} = \frac{8}{10}; \frac{8}{10} + 1 = \frac{9}{5}$$

Six-fold Symmetry: The Physical

$$\frac{360}{6} = 60; 360 - 60 - 60 = 240; \frac{240}{360} = \frac{2}{3}; \frac{2}{3} + 1 = \frac{5}{3}$$

Alternate Six-fold: The Physical

$$\frac{360}{6} = 60; 360 - 60 = 300; \frac{300}{360} = \frac{5}{6}; \frac{5}{6} + 1 = \frac{11}{6}$$

9/5: 5, 14, 23, 32,... and 1.8, 3.6, 5.4, 7.2,...
$a_n = 7.2n - 4$

5/3: 8, 13, 18, 23,... and 1.7, 3.3, 5, 6.7,...
$a_n = 3.3n + 3$

11/6: 6, 17, 28, 39,.. and 11/6, 11/3, 11/2, 22/3,...

$\pi + \phi = 3.141 + 1.618 = 4.759; 7 = (5+9)/2$
$\pi + e = 3.141 + 2.718 = 5.859; 9/5 = 1.8$

We Have Three Equations

$$x = \frac{36}{5}t - 4; y = \frac{33}{10}t + 3; z = 9t - 5; set\_t = 0$$
$$\frac{5x + 20}{36} = \frac{10y - 30}{33} = \frac{z + 5}{9}; \frac{5}{36}x - \frac{10}{33}y - \frac{1}{9}z + \frac{10}{11} = 0$$
$$\nabla f = \langle \frac{5}{36}, -\frac{10}{33}, -\frac{1}{9} \rangle$$

$a = 5/36; b = -10/33; c = \sqrt{(5/36)^2 = (10/33)^2} = 0.3328; d = -1/9$

$\tan\alpha = b/a; \alpha = 65.358°; \tan\beta = d/c; \beta = -18.46°$

$-65.35°/15°/hour = -4.3572hr; 24hr - 4.35hr = 19.6428h$

$RA: 19^h 38^m 34^s$

$DEC: -18°27'36" = my\_signal$

$RA: 19^h 22^m 24.64^s$

$DEC: -26°25'17.01" = SETI\_Wow!\_Signal$

Eliminating t In Our Three Equations

$t = \frac{5}{36}x + \frac{20}{36}; t = \frac{10}{33}y - \frac{30}{33}; t = \frac{1}{9}z + \frac{5}{9}$

$\frac{5}{36}x + \frac{20}{36} = \frac{10}{33}y - \frac{30}{33}$

$\frac{5}{36}x - \frac{10}{33}y + \frac{145}{99} = 0$

$\frac{10}{33}y - \frac{30}{33} = \frac{1}{9}z + \frac{5}{9}$

$\frac{10}{33}y - \frac{1}{9}z - \frac{145}{99} = 0$

$\frac{5}{36}x - \frac{20}{33}y + \frac{1}{9}z + \frac{290}{99} = 0$

$\nabla f = \langle \frac{5}{36}, -\frac{20}{33}, \frac{1}{9} \rangle$

$\sqrt{{5/36}^2 + {20/33}^2} = \sqrt{0.01929 + 0.3673} = 0.62176 \approx L = levinson's\_number$

0.62176 is the magnitude of the right ascension vector that points to the constellation Aquila.

Aquila

Sagittarius And Aquila

For the Sagittarius vector, we took the parameter, t, to be zero. If you instead eliminate t in the equations the vector points to an open cluster in Aquila, NGC 6738.

We have: <5/36,-20/33,1/9>

a=5/36, b=-20/33 d=1/9
c=sqrt((5/36)^2 +(20/33)^2)=sqrt(0.01929 + 0.3672)=sqrt(0.38659)

=0.62176

tan(alpha)=b/a=-48/11=-4.363636
alpha=-77 degrees
(-77 degrees)/(15 degrees)/hour =-5 hours
24 hours - 5 hours = 19 hours

tan(beta)=d/c=(1/9)/0.62176=0.1787
beta = arctan(0.1787) = +10 degrees

This is an object with right ascension 19 hours
And declination 10 degrees

This is in the constellation Aquila and is around the open cluster NGC 6738

NGC 6738 Open Cluster In Aquila

19h 62m 06.59s

+11° 37' 27.5"

Magnitude 8.3G

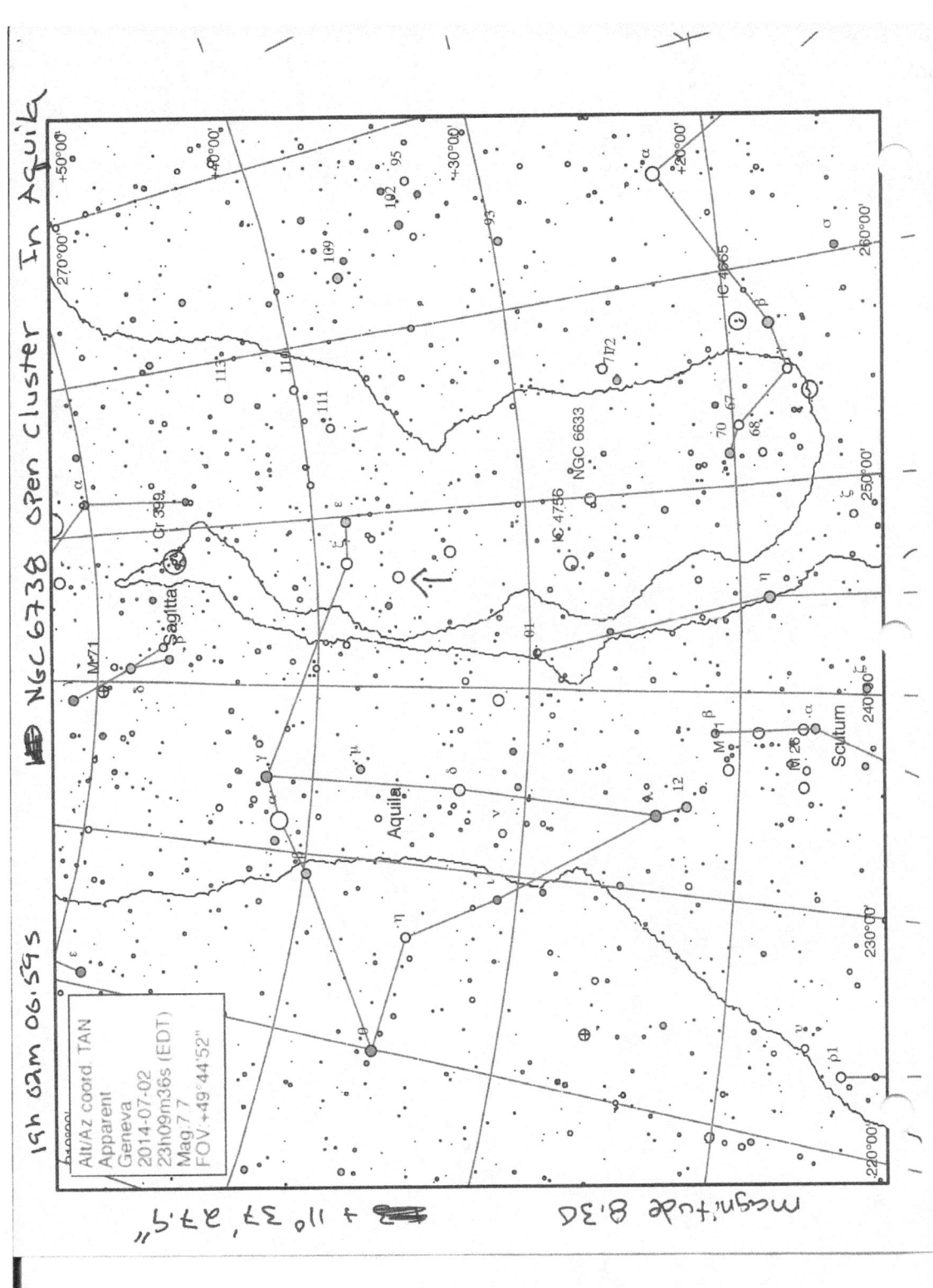

Next Contact

Another Projection For Extraterrestrial Contact

There is another way of calculating when the possible message from extraterrestrials in Sagittarius, the SETI Wow! Signal, will repeat it self. Our discovery of another message from the same place began with the Gypsy Shaman's hose collection of 15 hoses making us realize that 15 was important because the Earth rotates through 15 degrees in an hour, and 15 seconds lead to the dynamic integral, Manuel's Integral. The first message was on August 15, 1977. We noted that that 15 of August pointed to the Shaman's 15 hoses and the two sevens in 1977 add up to 14, which when added to the one in 19 is 15 as well, while the 9 is the nine of nine-fifths that we found in Nature from which we calculated a place in Sagittarius where the SETI Wow! Signal is. We decoded another message on around May 5 of 2013. The next message should then be on August 15, 2015, to line up the Shaman's fifteens. August is a good time to view the constellation Sagittarius from where I am in Southern California. Sagittarius has always been my favorite constellation, because it is in the center of the Galaxy, rich with globular, and open clusters that can be viewed with binoculars. Summer is when stargazing becomes exciting because you get both a rich sky and warm, uplifting weather. Also, August is the eighth month and our nine-fifths divided out is one point eight. The one takes you around a circle once, leaving point eight.

Ian Beardsley
Dec 24, 2013

Guess For Second Contact

It is my feeling that first contact with ET's was in Sagittarius (The Wow! Signal) because that is where my equations point when the parameter t, is zero. However, it is my feeling that second contact will be in Aquila, around NGC 6738, because that is where my equations point when I eliminate the parameter, t. I have indicated that one possible time for this second contact would be around August 15, 2015. Perhaps even on August 14, 15, and 16 of that same year.

Ian Beardsley
February 15, 2015

We Further Note,...

We can further note that, if we write another program called stlr.c, that the flux at golden ratio of earth orbit, for a star of golden ratio solar luminosities, that it is the golden ratio conjugate of the flux at earth orbit for the Sun. Close to it. And, that this is somewhere for that star that is close to the Mars orbital distance from the sun.

Ian Beardsley
March 08, 2015

The Program stelr.c

```c
#include<stdio.h>
#include<math.h>
int main(void)
{
float s, a, l, b, r, AU, N, root, number, answer, C, F;
printf("We determine the surface temperature of a planet.\n");
printf("What is the luminosity of the star in solar luminosities? ");
scanf("%f", &s);
printf("What is the albedo of the planet (0-1)?" );
scanf("%f", &a);
printf("What is the distance from the star in AU? ");
scanf("%f", &AU);
r=1.5E11*AU;
l=3.9E26*s;
b=l/(4*3.141*r*r);

N=(1-a)*b/(4*(5.67E-8));
root=sqrt(N);
number=sqrt(root);
answer=1.189*(number);

printf("The surface temperature of the planet is: %f K\n", answer);
C=answer-273;
F=(C*1.8)+32;
printf("That is %f C, or %f F", C, F);
printf("\n");
float joules;
joules=(3.9E26*s);
printf("The luminosity of the star in joules per second is: %.2fE25\n", joules/1E25);
float HZ;
HZ=sqrt(joules/3.9E26);
printf("The habitable zone of the star in AU is: %f\n", HZ);
printf("Flux at planet is %.2f times that at earth.\n", b/1370);
printf("That is %.2f Watts per square meter\n", b);
}
```

Running stelr.c For Golden Ratio And Its Conjugate

```
jharvard@appliance (~): cd dropbox/descubrir
bash: cd: dropbox/descubrir: No such file or directory
jharvard@appliance (~): cd Dropbox/descubrir
jharvard@appliance (~/Dropbox/descubrir): make stelr
clang -ggdb3 -O0 -std=c99 -Wall -Werror    stelr.c  -lcs50 -lm -o stelr
jharvard@appliance (~/Dropbox/descubrir): ./stelr
We determine the surface temperature of a planet.
What is the luminosity of the star in solar luminosities? 1
What is the albedo of the planet (0-1)?.3
What is the distance from the star in AU? 1
The surface temperature of the planet is: 303.727509 K
That is 30.727509 C, or 87.309517 F
The luminosity of the star in joules per second is: 39.00E25
The habitable zone of the star in AU is: 1.000000
Flux at planet is 1.01 times that at earth.
That is 1379.60 Watts per square meter
jharvard@appliance (~/Dropbox/descubrir): ./stelr
We determine the surface temperature of a planet.
What is the luminosity of the star in solar luminosities? 1.618
What is the albedo of the planet (0-1)?0.618
What is the distance from the star in AU? 1.618
The surface temperature of the planet is: 231.462616 K
That is -41.537384 C, or -42.767292 F
The luminosity of the star in joules per second is: 63.10E25
The habitable zone of the star in AU is: 1.272006
Flux at planet is 0.62 times that at earth.
That is 852.66 Watts per square meter
jharvard@appliance (~/Dropbox/descubrir):
```

Stellar Dot C Continued

By

Ian Beardsley

Copyright © 2015 by Ian Beardsley

I started out writing a program called *Discover* that searches for interconnections in the Universe. One of the routines in the program, called stellar.c calculated the surface temperature of a planet, given the luminosity of the star it orbits, the distance of the planet from the star, and the albedo of the planet. I then wrote a more sophisticated program called stelr.c that further determined the luminosity of the star in joules and gave the flux at the orbit of the planet, as well as the ratio of the flux at the planet to that at earth as produced by the sun. What I found was quite incredible. The theory behind making the calculations, and the programs, written in C, are given in appendix one. Here is what I found:

Notice running stellar.c for the golden ratio and its conjugate in terms of solar luminosities, earth orbit (AU) and albedo of the hypothetical planet gives near equivalence between the fahrenheit and centigrade scales for its surface temperature.

jharvard@appliance (~): cd Dropbox/descubrir
jharvard@appliance (~/Dropbox/descubrir): ./stellar
We determine the surface temperature of a planet.
What is the luminosity of the star in solar luminosities? 1.618
What is the albedo of the planet (0-1)?0.618
What is the distance from the star in AU? 1.618
The surface temperature of the planet is: 231.462616 K
That is -41.537384 C, or -42.767292 F
The luminosity of the star in joules per second is: 63.10E25
The habitable zone of the star in AU is: 1.272006
jharvard@appliance (~/Dropbox/descubrir):

We can further note that, if we write another program called stlr.c, that the flux at golden ratio of earth orbit, for a star of golden ratio solar luminosities, that it is the golden ratio conjugate of the flux at earth orbit for the Sun. Close to it. And, that this is somewhere for that star that is close to the Mars orbital distance from the sun.

jharvard@appliance (~/Dropbox/descubrir): ./stelr
We determine the surface temperature of a planet.
What is the luminosity of the star in solar luminosities? 1.618
What is the albedo of the planet (0-1)?0.618
What is the distance from the star in AU? 1.618
The surface temperature of the planet is: 231.462616 K
That is -41.537384 C, or -42.767292 F
The luminosity of the star in joules per second is: 63.10E25
The habitable zone of the star in AU is: 1.272006
Flux at planet is 0.62 times that at earth.
That is 852.66 Watts per square meter
jharvard@appliance (~/Dropbox/descubrir):

Our mystery star, with its mystery planet, connected to the Earth and the Sun by the golden ratio and its conjugate, has a luminosity of about the golden ratio conjugate times 10 to the 27. Twenty seven has the digits 2 and 7, which add to make nine and subtract to make 5, the 9 and 5 of the enigmatic nine-fifths. Our mystery star is spectral class F5 if on the main sequence as determined by calculating its absolute magnitude, which is +4.3. What region of the sky should we search to find it? In what constellation might it be?

Ian Beardsley March 15, 2015

$2.512^x = 1.618$

$x \log 2.512 = \log 1.618$

$(0.4)x = 0.209$

$x = 0.5$

$sun = +4.83$

$sun - 0.5 = 4.3$

$* = star$

$* = +4.3$

The foot-pound system was not derived from any relationship to nature that we know of. The Metric system was: one gram is the mass of a cube of water at STP one centimeter on each side. A centimeter is a hundredth of a meter, and a meter is a thousandth of a kilometer. One kilometer is one ten thousandth of the distance from the pole to the equator. Centigrade is derived such that there are 100 units between freezing and boiling points of water, freezing being zero and boiling, 100. Fahrenheit is not derived from anything natural: It stars at freezing of water at 32 degrees and who knows why? We do know the foot-pound system has earth gravity at 32 feet per second per second. Let us find the temperature where centigrade and Fahrenheit are the same:

F=(9/5)C + 32
F=C

C=(9/5)C=32
(25/25)C-(45/25)C=32
(-20/25)C=32

C=-32(25)/20=-800/20=-40 Degrees C

-40 degrees C = -40 degrees F

In earlier work, I have suggested that our different systems of measurement are connected not just to one another, but, to nature, even when the evolution of these systems had no such intent to do so.

What more can we say about the connection of Fahrenheit to Centigrade? We can say there are nine degrees of fahrenheit per five degrees of centigrade. The molar mass of gold to silver is nine to five, just like the ratio of the solar radius to the lunar orbital radius. The sun is gold in color and the moon is silver in color.

Pi and Phi

But is not nine-fifths a more dynamic number than what I have pointed out so far? Consider the golden ratio (denoted Φ called phi). And consider the ratio of the circumference of a circle to its diameter (denoted π called pi).

$$\Phi + \pi = 1.618 + 3.141 = 4.759$$

The four takes you around a circle 4 times, that is back to where you started, the fraction after 4, the 0.759, is the important part, it is what is left. Notice the seven is the average of nine and five and the 5 is the 5 of nine-fifths, and the nine is the nine of nine-fifths. We see that nine-fifths unifies the sum of pi and phi.

That which we are suggesting is, that there could be a star system in the universe which, through the golden ratio and its conjugate, and the scales of Fahrenheit and Centigrade, have its planet and star connected to the earth and the sun, and, that, in turn these star systems are connected to the relationship between pi and phi that is in nine-fifths.

Appendix 1

The Theory Behind stellar.c

As climate science is a new science, there are many models for the climate and I learned my climate science at MIT in a free online edX course. One can generate a basic model for climate with nothing more than high school algebra using nothing more than the temperature of the sun, the distance of the earth from the sun, and the earth's albedo, the percent of light it reflects back into space.

The luminosity of the sun is:

$$L_0 = 3.9 \times 10^{26} J/s$$

The separation between the earth and the sun is:

$$1.5 \times 10^{11} m$$

The solar luminosity at the earth is reduced by the inverse square law, so the solar constant is:

$$S_0 = \frac{3.9 \times 10^{26}}{4\pi(1.5 \times 10^{11})^2} = 1,370 Watts/meter^2$$

That is the effective energy hitting the earth per second per square meter. This radiation is equal to the temperature, T_e, to the fourth power by the steffan-bolzmann constant, sigma (σ). T_e can be called the effective temperature, the temperature entering the earth.

S_0 intercepts the earth disc, πr^2, and distributes itself over the entire earth surface, $4\pi r^2$, while 30% is reflected back into space due to the earth's albedo, a, which is equal to 0.3, so

$$\sigma T_e^4 = \frac{S_0}{4}(1-a)$$

$$(1-a)S_0 \frac{\pi r^2}{4\pi r^2}$$

But, just as the same amount of radiation that enters the system, leaves it, to have radiative equilibrium, the atmosphere radiates back to the surface so that the radiation from the atmosphere, σT_a^4 plus the radiation entering the earth, σT_e^4 is the radiation at the surface of the earth, σT_s^4. However,

$$\sigma T_a^4 = \sigma T_e^4$$

and we have:

$$\sigma T_s^4 = \sigma T_a^4 + \sigma T_e^4 = 2\sigma T_e^4$$

$$T_s = 2^{\frac{1}{4}} T_e$$

$$\sigma T_e^4 = \frac{S_0}{4}(1-a)$$

$$\sigma = 5.67 \times 10^{-8}$$

$$S_0 = 1,370$$

$$a = 0.3$$

$$\frac{1,370}{4}(0.7) = 239.75$$

$$T_e^4 = \frac{239.75}{5.67 \times 10^{-8}} = 4.228 \times 10^9$$

$$T_e = 255 Kelvin$$

So, for the temperature at the surface of the Earth:

$$T_s = 2^{\frac{1}{4}} T_e = 1.189(255) = 303 Kelvin$$

Let's convert that to degrees centigrade:

Degrees Centigrade = 303 - 273 = 30 degrees centigrade

And, let's convert that to Fahrenheit:

Degrees Fahrenheit = 30(9/5)+32=86 Degrees Fahrenheit

In reality this is warmer than the average annual temperature at the surface of the earth, but, in this model, we only considered radiative heat transfer and not convective heat transfer. In other words, there is cooling due to vaporization of water (the formation of clouds) and due to the condensation of water vapor into rain droplets (precipitation or the formation of rain).

The Program stellar.c

```c
#include<stdio.h>
#include<math.h>
int main(void)
{
float s, a, l, b, r, AU, N, root, number, answer, C, F;
printf("We determine the surface temperature of a planet.\n");
printf("What is the luminosity of the star in solar luminosities? ");
scanf("%f", &s);
printf("What is the albedo of the planet (0-1)?" );
scanf("%f", &a);
printf("What is the distance from the star in AU? ");
scanf("%f", &AU);
r=1.5E11*AU;
l=3.9E26*s;
b=l/(4*3.141*r*r);

N=(1-a)*b/(4*(5.67E-8));
root=sqrt(N);
number=sqrt(root);
answer=1.189*(number);

printf("The surface temperature of the planet is: %f K\n", answer);
C=answer-273;
F=(C*1.8)+32;
printf("That is %f C, or %f F", C, F);
printf("\n");
float joules;
joules=(3.9E26*s);
printf("The luminosity of the star in joules per second is: %.2fE25\n", joules/1E25);
float HZ;
HZ=sqrt(joules/3.9E26);
printf("The habitable zone of the star in AU is: %f\n", HZ);
}
```

The Program stelr.c

```c
#include<stdio.h>
#include<math.h>
int main(void)
{
float s, a, l, b, r, AU, N, root, number, answer, C, F;
printf("We determine the surface temperature of a planet.\n");
printf("What is the luminosity of the star in solar luminosities? ");
scanf("%f", &s);
printf("What is the albedo of the planet (0-1)?" );
scanf("%f", &a);
printf("What is the distance from the star in AU? ");
scanf("%f", &AU);
r=1.5E11*AU;
l=3.9E26*s;
b=l/(4*3.141*r*r);

N=(1-a)*b/(4*(5.67E-8));
root=sqrt(N);
number=sqrt(root);
answer=1.189*(number);

printf("The surface temperature of the planet is: %f K\n", answer);
C=answer-273;
F=(C*1.8)+32;
printf("That is %f C, or %f F", C, F);
printf("\n");
float joules;
joules=(3.9E26*s);
printf("The luminosity of the star in joules per second is: %.2fE25\n", joules/1E25);
float HZ;
HZ=sqrt(joules/3.9E26);
printf("The habitable zone of the star in AU is: %f\n", HZ);
printf("Flux at planet is %.2f times that at earth.\n", b/1370);
printf("That is %.2f Watts per square meter\n", b);
}
```

Appendix 2

Pi

The area of a circle is one half r time C, where r is its radius and C is its circumference. We immediately see that the ratio of the circumference to the diameter of a circle becomes important. It is the constant pi.

$$A = \frac{1}{2}rc$$
$$c = 2\pi r$$
$$A = \pi r^2$$

If we take a regular hexagon, which is a six-sided polygon with each side equal in length, and take each side equal to one, then if each side is one, so is the line drawn from each corner of the regular hexagon to its center, because it is made of equilateral triangles. And if we inscribe it in a circle, we can say the perimeter is close to the circumference of the circle and the line from each corner to the center (called a radius) is the same as the radius of the circle, then we have the ratio of the perimeter to the diameter is an approximation to pi and is 6/2 =3.00. If we increase the number of sides of the regular polygon, the perimeter comes closer and closer to the actual circumference of the circle and our value for pi becomes more accurate. If we increase the number of sides of the regular polygon enough times, we find that to three places after the decimal, pi is 3.141:

$$\pi = 3.141$$
$$A = \frac{1}{2}rc$$
$$D = 2r$$
$$c = 2\pi r$$
$$A = \pi r^2$$

The Golden Ratio

Let us draw a line and divide it such that the length of that line divided by the larger section is equal to the larger section divided by the smaller section. That ratio is The Golden Ratio, or phi:

$$\frac{a}{b} = \frac{b}{c}$$
$$a = b + c$$
$$c = a - b$$
$$a(a-b) = b^2$$
$$a^2 - ab = b^2$$
$$a^2 - ab - b^2 = 0$$
$$\left(\frac{a}{b}\right)^2 - \frac{a}{b} - 1 = 0$$
$$\left(\frac{a}{b}\right)^2 - \frac{a}{b} = 1$$
$$\left(\frac{a}{b}\right)^2 - \frac{a}{b} + \frac{1}{4} = \frac{5}{4}$$
$$\left(\frac{a}{b} - \frac{1}{2}\right)^2 = \frac{5}{4}$$
$$\frac{a}{b} = \frac{\sqrt{5}+1}{2} = 1.618...$$

Stellar Dot Python

By

Ian Beardsley

Copyright © 2015 by Ian Beardsley

In my book Stellar Dot C, I wrote a program called stellar dot c that is part of a bigger program called Discover. A story unfolded around that program. Now we write another program called stellar dot py around which another story unfolds. Discover was written as a series of programs that searches for interconnections in the universe and mathematics. Stellar Dot C was written in the language C. Stellar Dot PY is written in the language Python.

Sixty is of prime importance. An equilateral triangle has each of its angles with a measure of sixty degrees. The equilateral triangle is one of the three regular tessellators, of which the other two are the square and the regular hexagon. It is the face of the tetrahedron, which not only one of the five pythagorean, or platonic solids, but is the fundamental unit in Buckminster Fuller's Synergetics. It is quite convenient to divide an hour into 60 minutes, and a minute into 60 seconds, because of the divisibility of 60 into the 360 degrees of a circle, 6 times.

Carl Munck in his brilliant work, The Code, has made an extraordinary discovery. He has said the way of measuring the earth in ancient times was different, but that found that it was related to our modern methods. He takes Stonehenge and says there were originally 60 stones around the perimeter. He takes the 360 degrees in a circle and multiplies them by 60 to get the number 21,600.

Here is the incredible thing he found: The latitude of Stonehenge in our modern way of measuring the earth is 51 degrees 10 minutes 42.35 seconds. Watch the incredible thing he found, that happens:

(51)(10)(42.35)=21,598.5

Which is very close to the 21,600 he got for the latitude of Stonehenge in what he surmises they used to measure the earth. He next built the ancient grid and shows how it is related to most of the ancient monuments.

He took the Great Pyramid of Giza as The Prime Meridian. I have set out to write a program that calculates the surface temperature of a planet given the luminosity of the star that it orbits, the distance of the planet from the star, and the albedo of the planet, which is the fraction of light reaching the planet that is reflected back into space. Let us first present the theory used to determine the surface temperature of the planet, then present the program in python called stellar dot python (stellar.py). Then, we will run the program for some interesting values and we will see that when we do that, something interesting happens.

Since Munck found a reference location in space for the ancient grid, I am looking for reference for location in time. He said that Egypt's Great Pyramid was the prime meridian in the ancient grid. Here is what I have done. The Sothic cycle of the Egyptian Calendar is 1,460 Julian years. A Julian year is 365.25 days. A star rises 4 minutes earlier each night; that is 240 seconds. 60 is not just important because there were 60 stones around Stonehenge, but because there are 60 minutes in an hour and 60 seconds in a minute and 60 degrees are the angles in an equilateral triangle, a shape particularly important in

Buckminster Fuller's Synergetics. Watch what happens if I multiply the Sothic cycle by 60 and divide by the 240 seconds earlier that a star rises each night:

$(1,460)(60)/240 = 365$ days

There are 1,461 days in a Sothic cycle in Egyptian years. An Egyptian year is 365 days. It is important that I relate everything to 240 seconds, because in my work ET Conjecture, I have shown the Sothic cycle related to the square root of two over two, which is the ratio of the side of a square to its radius, to the golden ratio conjugate, six-fold symmetry of the physical universe (five fold is biological), the 240 seconds a star rises earlier each night, all this by the standard reference for concert pitch A440, that the oboe sounds before the orchestra performs to tune their instruments to one frequency. With this, I think we can find where time is zero in Munck's grid. Time equal to zero just may be the hypothesized beginning of the Egyptian calendar, which is 4242 BC, four Sothic Cycles ago.

If we use the Sothic cycle in Julian years (365.25 days) we get a Julian year:

$(1,461)(60)/240 = 365.25$

The Theory Behind Stellar Dot PY

As climate science is a new science, there are many models for the climate and I learned my climate science at MIT in a free online edX course. One can generate a basic model for climate with nothing more than high school algebra using nothing more than the temperature of the sun, the distance of the earth from the sun, and the earth's albedo, the percent of light it reflects back into space.

The luminosity of the sun is:

$$L_0 = 3.9 \times 10^{26} \, J/s$$

The separation between the earth and the sun is:

$$1.5 \times 10^{11} \, m$$

The solar luminosity at the earth is reduced by the inverse square law, so the solar constant is:

$$S_0 = \frac{3.9 \times 10^{26}}{4\pi(1.5 \times 10^{11})^2} = 1{,}370 \, Watts/meter^2$$

That is the effective energy hitting the earth per second per square meter. This radiation is equal to the temperature, T_e, to the fourth power by the steffan-bolzmann constant, sigma (σ). T_e can be called the effective temperature, the temperature entering the earth.

S_0 intercepts the earth disc, πr^2, and distributes itself over the entire earth surface, $4\pi r^2$, while 30% is reflected back into space due to the earth's albedo, a, which is equal to 0.3, so

$$\sigma T_e^4 = \frac{S_0}{4}(1-a)$$

$$(1-a)S_0 \frac{\pi r^2}{4\pi r^2}$$

But, just as the same amount of radiation that enters the system, leaves it, to have radiative equilibrium, the atmosphere radiates back to the surface so that the radiation from the atmosphere, σT_a^4 plus the radiation entering the earth, σT_e^4 is the radiation at the surface of the earth, σT_s^4. However,

$$\sigma T_a^4 = \sigma T_e^4$$

and we have:

$$\sigma T_s^4 = \sigma T_a^4 + \sigma T_e^4 = 2\sigma T_e^4$$

$$T_s = 2^{\frac{1}{4}} T_e$$

$$\sigma T_e^4 = \frac{S_0}{4}(1-a)$$

$$\sigma = 5.67 \times 10^{-8}$$

$$S_0 = 1{,}370$$

$$a = 0.3$$

$$\frac{1{,}370}{4}(0.7) = 239.75$$

$$T_e^4 = \frac{239.75}{5.67 \times 10^{-8}} = 4.228 \times 10^9$$

$$T_e = 255 \, Kelvin$$

So, for the temperature at the surface of the Earth:

$$T_s = 2^{\frac{1}{4}} T_e = 1.189(255) = 303 \, Kelvin$$

Let's convert that to degrees centigrade:

Degrees Centigrade = 303 - 273 = 30 degrees centigrade

And, let's convert that to Fahrenheit:

Degrees Fahrenheit = 30(9/5)+32=86 Degrees Fahrenheit

In reality this is warmer than the average annual temperature at the surface of the earth, but, in this model, we only considered radiative heat transfer and not convective heat transfer. In other words, there is cooling due to vaporization of water (the formation of clouds) and due to the condensation of water vapor into rain droplets (precipitation or the formation of rain).

The Program stellar.py

```
print("We determine the surface temperature of a planet.")
s=float(raw_input("Enter stellar luminosity in solar luminosities: "))
a=float(raw_input("What is planet albedo (0-1)?: "))
au=float(raw_input("What is the distance from star in AU?: "))
r=(1.5)*(10**11)*au
l=(3.9)*(10**26)*s
b=l/((4.0)*(3.141)*(r**2))
N=((1-a)*b)/(4.0*((5.67)*(10**(-8))))
root=N**(1.0/2.0)
number=root**(1.0/2.0)
answer=1.189*number
print("The surface temperature of the planet is: "+str(answer)+"K")
C=answer-273
F=(9.0/5.0)*C + 32
print("That is " +str(C)+"C")
print("Which is " +str(F)+"F")
joules=3.9*(10**26)*s/1E25
lum=(3.9E26)*s
print("luminosity of star in joules per sec: "+str(joules)+"E25")
HZ=((lum/(3.9*10**26)))**(1.0/2.0)
print("The habitable zone is: "+str(HZ))
flux=b/1370.0
print("Flux at planet is "+str(flux)+" times that at earth")
print("That is " +str(b)+ " watts per square meter")
```

Running stellar.py

In [1] %run " /Users/ianbeardsley/Desktop/6.00x Files/stellar.py"

We determine the surface temperature of a planet.

Enter stellar luminosity in solar luminosities: 60

What is the albedo (0-1)?: 0.605

What is the distance from star in AU?: 7.2
The surface temperature of the planet is 273.042637637K
That is 0.0426376371411C
Which is 32.0767477469F
luminosity of star in joules per sec: 2340.0E25
The habitable zone is 7.74596669241
Flux at planet is 1.16552032543 times that at earth
That is 1596.76284583

In [2]

Notice running for the key numbers 60 (60 sec in minute, 60 min in hour, 60 degrees in angles of equilateral triangle), 0.605 ~0.618=golden_ratio conjugate, 7.2 (the earth precesses through one degree every 72 years, Venus is at 0.72 AU from sun, 72 is room temperature), that we get close to the freezing temperature of water (0 degrees centigrade, 32 degrees fahrenheit, 273 degrees kelvin).

We find the absolute magnitude of the star around which the planet orbits, and from that determine its spectral class if on the main sequence.

x log 2.512 = log 60

0.4x = 1.778

x= +4.445

sun = +4.83

+4.83 - 4.445 = +0.385

* = star
* = +0.385

The absolute magnitude of the star is 0.385 which means on the main sequence it is about A5V, a blue star.

Running stellar.py For Earth And Sun At Golden Ratio Conjugate

In [1] %run "/Users/ianbeardsley/Desktop/6.00.1x Files/stellar.py"
We determine the surface temperature of a planet.

Enter stellar luminosity in solar luminosities: 1

What is planet albedo (0-1)?: 0.6

What is the distance from star in AU?: 1

The surface temperature of the planet is 264.073383631
That is -8.92661636938C
Which is 15.9320905351F
luminosity of star in joules per sec: 39.0E25
The habitable zone is 1
Flux at planet is 1.00700956117
That is 1379.6030988 watts per square meter

In [2]

Notice if we increase the albedo of the earth to the golden ratio conjugate (0.6), that is the earth reflects 60% of the light it receives back into space instead of what it actually reflects back into space which is 30%, we get that the surface temperature in Fahrenheit is close to what the actual average surface temperature is in degrees centigrade which is 15. Notice that in degrees centigrade, the surface temperature would be close to -9 degrees centigrade. 9 and 15 are bound to one another in my theories in the my works ET Conjecture, The Program Discover, and The Genesis Project.

I started out writing a program called *Discover* that searches for interconnections in the Universe. One of the routines in the program, called stellar.c calculated the surface temperature of a planet, given the luminosity of the star it orbits, the distance of the planet from the star, and the albedo of the planet. I then wrote a more sophisticated program called stelr.c that further determined the luminosity of the star in joules and gave the flux at the orbit of the planet, as well as the ratio of the flux at the planet to that at earth as produced by the sun. What I found was quite incredible. The theory behind making the calculations, and the programs, written in C, are given in appendix one. Here is what I found:

Notice running stellar.c for the golden ratio and its conjugate in terms of solar luminosities, earth orbit (AU) and albedo of the hypothetical planet gives near equivalence between the fahrenheit and centigrade scales for its surface temperature.

```
jharvard@appliance (~): cd Dropbox/descubrir
jharvard@appliance (~/Dropbox/descubrir): ./stellar
We determine the surface temperature of a planet.
What is the luminosity of the star in solar luminosities? 1.618
What is the albedo of the planet (0-1)?0.618
What is the distance from the star in AU? 1.618
The surface temperature of the planet is: 231.462616 K
That is -41.537384 C, or -42.767292 F
The luminosity of the star in joules per second is: 63.10E25
The habitable zone of the star in AU is: 1.272006
jharvard@appliance (~/Dropbox/descubrir):
```

We can further note that, if we write another program called stlr.c, that the flux at golden ratio of earth orbit, for a star of golden ratio solar luminosities, that it is the golden ratio conjugate of the flux at earth orbit for the Sun. Close to it. And, that this is somewhere for that star that is close to the Mars orbital distance from the sun.

```
jharvard@appliance (~/Dropbox/descubrir): ./stelr
We determine the surface temperature of a planet.
What is the luminosity of the star in solar luminosities? 1.618
What is the albedo of the planet (0-1)?0.618
What is the distance from the star in AU? 1.618
The surface temperature of the planet is: 231.462616 K
That is -41.537384 C, or -42.767292 F
The luminosity of the star in joules per second is: 63.10E25
The habitable zone of the star in AU is: 1.272006
Flux at planet is 0.62 times that at earth.
That is 852.66 Watts per square meter
jharvard@appliance (~/Dropbox/descubrir):
```

Running stellar.py

In [1] %run " /Users/ianbeardsley/Desktop/6.00x Files/stellar.py"

We determine the surface temperature of a planet.

Enter stellar luminosity in solar luminosities: 60

What is the albedo (0-1)?: 0.605

What is the distance from star in AU?: 7.2
The surface temperature of the planet is 273.042637637K
That is 0.0426376371411C
Which is 32.0767477469F
luminosity of star in joules per sec: 2340.0E25
The habitable zone is 7.74596669241
Flux at planet is 1.16552032543 times that at earth
That is 1596.76284583

In [2]

Notice running for the key numbers 60 (60 sec in minute, 60 min in hour, 60 degrees in angles of equilateral triangle), 0.605 ~0.618=golden_ratio conjugate, 7.2 (the earth precesses through one degree every 72 years, Venus is at 0.72 AU from sun, 72 is room temperature), that we get close to the freezing temperature of water (0 degrees centigrade, 32 degrees fahrenheit, 273 degrees kelvin).

My program models a planet without convection as a factor for planetary cooling. When we run this program for three sets of data involving the golden ratio, among other things, we get three results of profound significance, which seem to be connected to a Project Genesis wherein we try to understand how to create and maintain life sustaining planets.

Ian Beardsley
March 23, 2015

Albedo

Albedo is a function of surface reflectivity and atmospheric reflectivity. Atmospheric albedo seems to play the primary role in the overall albedo of a planet. Albedo is the percent of light incident to a surface that is reflected back into space. It has a value ranging from zero to one inclusive. Zero is a black surface absorbing all incident light and one is a white surface reflecting all incident light back into space. Albedo plays a dominant role in the climate of a planet. Let us see if we can find a relationship between composition of a planet and its albedo if not in its distance from the star it orbits and its albedo, even a relationship between its albedo and orbital number, in that albedo could be a function of distance from the star a planet orbits because composition seems to be a function of distance of a planet from the star it orbits. As in the inner planets are solid, or terrestrial, and the outer planets are gas giants. There may be an analogue to the Titius-Bode rule for planetary distribution, but for albedo with respect to planetary number. The inner planets are dominantly CO_2, Nitrogen, Oxygen, and water vapor, the outer planets, hydrogen and helium.

1. Mercury albedo of 0.06 composition 95% CO_2
2. Venus albedo of 0.75 composition clouds of sulfuric acid
3. Earth albedo of 0.30 composition Nitrogen, Oxygen, $H20$ or water vapor
4. Mars albedo of 0.29 composition CO_2
5. Asteroids
6. Jupiter albedo of 0.53 composition hydrogen and helium
7. Saturn albedo of 0.47 composition hydrogen and helium
8. Uranus albedo of 0.51 composition hydrogen, helium, methane
9. Neptune albedo of 0.41 composition of hydrogen and helium

We see the outer gas giant, which are composed chiefly of hydrogen and helium have albedos around 50%. Earth and Mars, the two planets in the habitable zone, are about the same (30%).

Go to the next page for a graph of albedo to planetary number.

mercury	0.06
venus	0.75
earth	0.3
mars	0.29
asteroids	
jupiter	0.52
saturn	0.47
uranus	0.51
neptune	0.41

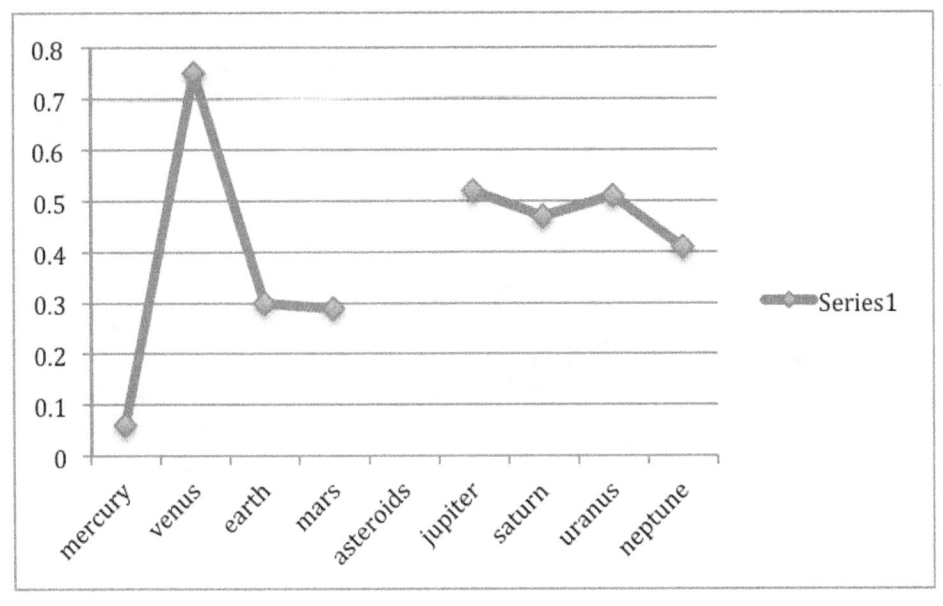

The average for the albedo of the inner planets is:
(0.06+0.75+0.3+0.29)/4 = 0.35
This is close to the albedo of the habitable planets Earth and Mars.

The average for the albedo of the outer planets is:

(0.52+0.47+0.51+0.41)/4 + 0.4775 ~0.48
This says the outer planets are all close to 0.48~0.5

All this also says, if the planet is solid and habitable it probably has an albedo of around 0.3, otherwise it is an outer gaseous planet and probably has an albedo of around 0.5.

The Venus Enigma of Stellar Dot Python

The high albedo of Venus is due to clouds of sulfuric acid. Since it has no oceans and thereby does not cool by convection, if placed at earth orbit it would fit the scenario for a planet at 1 AU, orbiting a star of one stellar luminosity, to have the temperature of 15F talked about in stellar dot python, that is if it has an albedo of golden ratio conjugate (0.6).

Manuel's Number And Nine-fifths Connected to Climate

The solar luminosity is:

$$L_0 = 3.9 \times 10^{26} J/s$$

The average distance of the Earth from the Sun is:

$$1.5 \times 10^{11} m$$

Therefore the solar constant is:

$$S_0 = \frac{3.9 \times 10^{26}}{4\pi(1.5 \times 10^{11})^2} = 1,370 watts/meter^2$$

That is the amount of energy per second per square meter hitting the Earth.

The radiation, F, is proportional to the temperature, T to the fourth power, and equal by the Stefan-Boltzman constant, sigma:

$$F = \sigma T^4$$
$$\sigma = 5.67 \times 10^{-8} Wm^{-2}K^{-4}$$

The planetary albedo, a, is the amount of radiation from the Sun that the Earth reflects back into space which is 30%. Therefore a=0.3 is the planetary albedo. Therefore the Earth receives 70% of the Sun's light, or, in other words:

$$\frac{S_0(1-a)\pi r^2}{4\pi r^2} = \frac{S_0}{4}(1-a) = \sigma T^4$$
$$T = 255K = -18°C$$

That is the temperature the Earth would be if it had no atmosphere, minus eighteen degrees centigrade. The observed average temperature is:

$$T = 15°C$$

Fifteen degrees centigrade. The without an atmosphere temperature, the minus 18 degrees centigrade, is the one and eight of 1.8 that is the Nine-Fifths upon which we have built our Gypsy Cosmology. The fifteen Degrees is the Gypsy's 15.

F=1.8(-18)+32=-0.4 degrees F

This gives the interesting result of about 0 degrees Fahrenheit.

The Case For Mars

We have found the interesting models for climate of a planet -40 degrees C = -40 degrees F, 0 degrees F, 0 degrees C = 32 degrees F, 15 degrees F = average annual earth temperature, and left to find is the interesting case of ten solar luminosities (we use a base ten counting system) with albedo of golden ratio conjugate and mars orbit = 1.523 AU and we get about 103 degrees C which is close to 100 degrees C at which water boils that is used to define the centigrade system. The habitable zone is close to pi in astronomical units. We are finding the cosmos is structured around our system of units of measurement if we look at star systems.

```
jharvard@appliance (~): cd Dropbox/descubrir
jharvard@appliance (~/Dropbox/descubrir): ./stelr
We determine the surface temperature of a planet.
What is the luminosity of the star in solar luminosities? 10
What is the albedo of the planet (0-1)?0.618
What is the distance from the star in AU? 1.523
The surface temperature of the planet is: 376.162537 K
That is 103.162537 C, or 217.692566 F
The luminosity of the star in joules per second is: 390.00E25
The habitable zone of the star in AU is: 3.162278
Flux at planet is 4.34 times that at earth.
That is 5947.77 Watts per square meter
jharvard@appliance (~/Dropbox/descubrir):
```

An Extraterrestrial Beacon And Manuel's Third Integral

By Ian Beardsley

Copyright © 2015 by Ian Beardsley

It has always been thought that an extraterrestrial message would be mathematical. Are the anomalous Fast Radio Bursts (FRBs) an extraterrestrial beacon encoding Manuel's Third Integral that I discovered when doing research that lead me to the other theoretical ET signal, the Wow Signal in the constellation Sagittarius? Here I present the integral and how it is connected to the FRBs. To learn of all the research, read my work "SETI: Another Signal in Sagittarius" by Ian Beardsley or my work "ET Conjecture" (Red 01) with the title page "This Is 440" at the opening. As well read my work "The Program Discover" and "The Genesis Project".

Ian Beardsley
April 2, 2015

ET Conjecture: http://issuu.com/discover4/docs/red_01

FRBs and Manuel's Third Integral

The occurrence of FRBs (Fast Radio Bursts) from outside the Galaxy that can be explained by no known natural process I find have their source to be in an integral that was brought to me by a Gypsy Shaman, Manuel. There are three Manuel Integrals, and my first one has been part of an overall theory that does not just propose extraterrestrial contact in 2015, but tells us where to look. Manuel's third integral is derived from the nine fifths that is the ratio of the molar masses of gold to silver and is in the ratio of the solar radius to the lunar orbital radius, the sun gold in color, the moon silver. It also uses earth gravity rounded to the nearest ten, and the 15 that describes the amount of degrees through which the earth passes in one hour. The New Scientist reports on March 31, 2015 that the time between the beginning of the first burst and the end of the last burst is a multiple of 187.5. This happens to be the same as Manuel's Third Integral, the decimal part beyond 9 kilometers:

We consider the earth equation:

(y) = 9x-5
v=9t-5
g=980 cm/s/s
9(980)=8,820 cm/s/s

5=9t
(t)=5/9

(v_0) = (8,820 cm/s/s)(5/9) =4,900 cm/s

dx = (8,820 cm/s/s)t dt – (4,900 cm/s) dt

$$\int_0^{15} 8,820 cm/s/s\, dt - \int_0^{15} 4,900 cm/s\, dt = 9.1875 km$$

The Integral has the 9 of nine-fifths, the five, the seven, which is the average of 9 and 5 and the one and eight of the 1.8 that is 9/5 divided out. We call this Manuel's Third Integral.

Ian Beardsley
April 1, 2014

Once you realize nine-fifths is not just at the crux of Gold and Silver, Pi and the Golden Ratio, Pi and Euler's Number e, the five-fold symmetry that is typical of life, Jupiter and Saturn, Sun and Moon, it is not long before you realize its compliment is 5/3 and that you form the sequences:

(For 9/5) 5, 14, 23, 32,... and 1.8, 3.6, 5.4, 7.2,...
(For 5/3) 8, 13, 18, 23,... and 1.7, 3.3, 5, 6.7,...

For which you get:

$7.2n - 4 = a\_n$ and $(a\_n) = 3 + 3.3n$ respectively.

In the latter, letting the 3.3 be Earth Gravities rounded to the nearest ten (980), we have:
(v) = 2,940 cm/s + (3,234 cm/s/s)t
This is the differential equation:

(dx) = (2,940 cm/s)dt + (3,234 cm/s/s)t dt

$$\int_0^{15}(2,940 cm/s)dt + \int_0^{15}(3,234 cm/s/s)t dt = 4.07925 km$$ (Manuel's Integral)

15 seconds because there are 15 degrees in an hour of right ascension. The factor of one fourth enters because the kilometer is defined by the distance from the pole to the equator, not by the circumference of the Earth. Notice the 0.07925 has the nine and five of nine-fifths, the average of nine-fifths and the 2 used to make it.

Mach 1 = 768 mph =1,235 km/hour
That is mach 1 in dry air at 20 degrees C (68 degrees F) at sea level.
If we write, where 1,235 km/hr (mach 1) = 0.343 km/s, then:
34,300 cm/s =2,940 cm/s + (3234 cm/s/s)t
and
t=9.696969697 seconds = 9 23/33 s = 320/33 seconds ~ 9.7 seconds
So, the Uranus equation is a time of 9.7 seconds to reach mach 1. Putting that time in the integral:
(x) = (2,940)(320/33) + 1/2(3234(320/33)^2 = 180,557 cm 1.80557 km ~ 1.8km
Thus we see Manuel's Integral reaches mach one in about 9.7 seconds after traveling a distance of about 1.8 kilometers. Let's convert that to miles:
(1.8km)(0.621371mi/km)=1.118478 miles
Manuel's Integral reaches mach one in 1.8 kilometers, which is the amount of kilometers in a mile and is the 9/5 that occurs in Nature and the Universe, not to mention that it unifies pi and golden ratio and pi and Euler's number e. It is one compact statement that embodies everything and connects it to Earth Gravity. – Ian Beardsley, August 15, 2013

Manuel's Second Integral

Earth gravity (g) is 9.81 m/s/s
This is close to 9.80 m/s/s
Indeed if rounded it to one place after the decimal, it would be 9.8 m/s/s
This value when converting to cm/s/s gives g = 980 cm/s/s
There may be good reason to write it like this (which is rounding it to the nearest ten) because we see in our research that it is fruitful not mention that it provides a nice form for the value if we want to create a new system of units both with a zero at the end for the value and that is connected to nature, which it is, in Manuel's integral. Also the nine is the nine in the nine-fifths connected to nature and mathematical constants, as we have shown in our research, and the eight is the 0.8 in the 1.8 that is nine fifths, the fraction around a circumference of a circle that is nine-fifths of a circumference.

Let us consider the Neptune Equation:

7.2x −4 = y

Let 7.2 be Earth Gravities:

(v) = 7.2t −4
(dx/dt) = 7.2t −4
(dx) = 7.2t dt − 4dt
(7.2)(980 cm/s/s) = 7,056 cm/s/s

v=4=7.2t
t=(5/9)
(7,056)(5/9)=v_0 = 3,920 so,

(dx) = 7,056 cm/s/s t dt −3,920 cm/s dt

$$\int_0^{15} 7,056 cm/s/s t dt - \int_0^{15} 3,920 cm/s dt = 7.35 km = \frac{147}{20} km$$

We call this Manuel's Second Integral

Ian Beardsley
November 26, 2013

The Author

www.ingramcontent.com/pod-product-compliance
Lightning Source LLC
Chambersburg PA
CBHW080923170526
45158CB00008B/2211